ENERGY SCIENCE, ENGINEERING AND TECHNOLOGY

THE KEYSTONE XL PIPELINE PROJECT: PROPOSALS AND CONSIDERATIONS

ENERGY SCIENCE, ENGINEERING AND TECHNOLOGY

Additional books in this series can be found on Nova's website under the Series tab.

Additional E-books in this series can be found on Nova's website under the E-book tab.

ENERGY POLICIES, POLITICS AND PRICES

Additional books in this series can be found on Nova's website under the Series tab.

Additional E-books in this series can be found on Nova's website under the E-book tab.

ENERGY SCIENCE, ENGINEERING AND TECHNOLOGY

THE KEYSTONE XL PIPELINE PROJECT: PROPOSALS AND CONSIDERATIONS

GREGORY B. FLYNN
AND
DEBRA BURNETT
EDITORS

Nova Science Publishers, Inc.
New York

For permission to use material from this book please contact us:
Telephone 631-231-7269; Fax 631-231-8175
Web Site: http://www.novapublishers.com

Additional color graphics may be available in the e-book version of this book.

Library of Congress Cataloging-in-Publication Data

ISBN 978-1-62081-221-1

Published by Nova Science Publishers, Inc. † New York

CONTENTS

PREFACE

TransCanada (a Canadian company) applied to the U.S. Department of State for a permit to cross the U.S.-Canada International border with the Keystone XL pipeline project in 2008. If constructed, the pipeline would carry crude oil produced from the oil sands regions of Alberta, Canada, to U.S. Gulf Coast refineries. Because the pipeline would connect the United States with a foreign country, it requires a Presidential Permit issued by the State Department. Issuance of a Presidential Permit requires a finding that the project would serve the "national interest." This book describes the Keystone XL pipeline proposal and the process required for federal approval. It summarizes key arguments for and against the pipeline put forth by the pipeline's developers, federal agencies, environmental groups, and other stakeholders. Also discussed is the constitutional basis for the State Department's authority to issue a Presidential Permit, and opponents' possible challenges to this authority.

Chapter 1- In 2008, Canadian pipeline company TransCanada filed an application with the U.S. Department of State to build the Keystone XL pipeline, which would transport crude oil from the oil sands region of Alberta, Canada, to refineries on the U.S. Gulf Coast. Keystone XL would ultimately have the capacity to transport 830,000 barrels per day, delivering crude oil to the market hub at Cushing, OK, and further to points in Texas. TransCanada plans to build a pipeline spur so that oil from the Bakken formation in Montana and North Dakota can also be carried on Keystone XL.

As a facility connecting the United States with a foreign country, the pipeline requires a Presidential Permit from the State Department. In evaluating such a permit application, after consultation with other relevant federal agencies and public input, the department must determine whether a proposal is in the "national interest." This determination considers the project's potential effects on the environment, economy, energy security, foreign policy, and other factors. Pursuant to the National Environmental Policy Act, the State Department considered potential environmental impacts of the proposed Keystone XL project in a final Environmental Impact Statement (EIS) issued on August 26, 2011. A wide range of public comments both for and against the pipeline were received during a subsequent 90-day review period. The State Department noted, in particular, concerns about the pipeline's route through the Sand Hills region of Nebraska, an extensive sand dune formation with highly porous soil and shallow groundwater.

Chapter 2- In 2008, TransCanada Corp. applied for a presidential permit from the State Department to construct and operate an oil pipeline across the U.S.-Canada border in a project known as Keystone XL. The Keystone XL pipeline would transport oil produced from oil sands in Alberta, Canada, to Gulf Coast refineries. The permit application was subjected to review by the State Department pursuant to executive branch authority over cross-border pipeline facilities as articulated in Executive Order 13337.

After several phases of review, on November 10, 2011, the State Department announced that it would seek additional information about alternative pipeline routes before it could move forward with a national interest determination. In response, several pieces of legislation were introduced, including Title V of the Temporary Payroll Tax Cut Continuation Act of 2011. Title V dictated that President must grant the Keystone XL pipeline permit within 60 days of the law's enactment, unless the President determined that the pipeline is not in the national interest. If the President did not make a national interest determination and took no action to grant the permit, then the law provided that the permit "shall be in effect by operation of law." The Temporary Payroll Tax Cut Continuation Act of 2011 (P.L. 112-78), including Title V addressing the Keystone XL permit, was enacted on December 23, 2011.

Chapter 3- The proposed Keystone XL Project consists of a crude oil pipeline and related facilities that would primarily be used to transport Western Canadian Sedimentary Basin crude oil from an oil supply hub near Hardisty, Alberta Canada to delivery points in Oklahoma and Texas. The proposed Project would also be capable of transporting U.S. crude oil to those delivery points. The U.S. portion of the pipeline would begin near Morgan, Montana at the international border of the United States and extend to delivery points in Nederland and Moore Junction, Texas. There would also be a delivery point at Cushing, Oklahoma. These three delivery points would provide access to many other U.S. pipeline systems and terminals, including pipelines to refineries in the U.S. Gulf Coast region. Market conditions, not the operator of the pipeline, would determine the refining locations of the crude oil.

The proposed Keystone XL pipeline would consist of approximately 1,711 miles of new 36-inch-diameter pipeline, with approximately 327 miles of pipeline in Canada and 1,384 miles in the U.S. Figure ES-3 depicts the three segments of the proposed Project in the U.S. As noted in that illustration, the proposed Project would connect to the northern and southern ends of the existing Cushing Extension of the Keystone Oil Pipeline System.

Chapter 4- The authors wanted to talk today about the announcement that they have made. The Department of State recommended to President Obama that the presidential permit for the proposed Keystone XL pipeline be denied and that at this time the TransCanada Keystone XL pipeline be determined to not serve the national interest. The President agreed with the Department's recommendation. This recommendation was predicated on the fact that the Temporary Payroll Tax Cut Continuation Act that was passed in December does not provide sufficient time to obtain the information that the authors think is necessary to assess whether the project, in its current state, is in the national interest.

Chapter 5- The Keystone XL project is a proposed seven billion dollar, 1,700-mile pipeline that TransCanada Keystone Pipeline LP applied to build and operate. It would transport up to 830,000 barrels per day of crude oil from Alberta, Canada to refineries in the U.S. Gulf Coast Region. Most of the oil would originate from the oil sands of Alberta, although as much as 100,000 barrels per day might have come from the Bakken formation in Montana and North Dakota and up to 150,000 barrels per day could have been transported from Cushing, Oklahoma to the Gulf Coast. The U.S. Department of State received the application for this project in September 2008. The authors began a thorough, rigorous and transparent process to determine whether issuance of a Presidential Permit for this pipeline was in the national interest. Congress passed the Temporary Payroll Tax Cut Continuation Act of 2011, which required a determination by the President, within 60 days, of whether the Keystone XL pipeline project would serve the national interest. On January 18, 2012 the Department of State recommended to the President that the application for a Presidential Permit be denied, due to insufficient time to conduct the necessary analysis, and the President accepted the authors' recommendation and determined that the Keystone XL pipeline project, as presented and analyzed at that time, would not serve the national interest.

In: The Keystone XL Pipeline Project
Editors: G. B. Flynn and D. Burnett
ISBN: 978-1- 62081-221-1

Chapter 1

KEYSTONE XL PIPELINE PROJECT: KEY ISSUES*

Paul W. Parfomak, Linda Luther and Adam Vann

SUMMARY

In 2008, Canadian pipeline company TransCanada filed an application with the U.S. Department of State to build the Keystone XL pipeline, which would transport crude oil from the oil sands region of Alberta, Canada, to refineries on the U.S. Gulf Coast. Keystone XL would ultimately have the capacity to transport 830,000 barrels per day, delivering crude oil to the market hub at Cushing, OK, and further to points in Texas. TransCanada plans to build a pipeline spur so that oil from the Bakken formation in Montana and North Dakota can also be carried on Keystone XL.

As a facility connecting the United States with a foreign country, the pipeline requires a Presidential Permit from the State Department. In evaluating such a permit application, after consultation with other relevant federal agencies and public input, the department must determine whether a proposal is in the "national interest." This determination considers the project's potential effects on the environment, economy, energy security, foreign policy, and other factors. Pursuant to the National Environmental Policy Act, the State Department considered potential environmental impacts

---

* This is an edited, reformatted and augmented version of the Congressional Research Service publication, R41668, dated January 26, 2012.

of the proposed Keystone XL project in a final Environmental Impact Statement (EIS) issued on August 26, 2011. A wide range of public comments both for and against the pipeline were received during a subsequent 90-day review period. The State Department noted, in particular, concerns about the pipeline's route through the Sand Hills region of Nebraska, an extensive sand dune formation with highly porous soil and shallow groundwater.

On November 10, 2011, in response to concerns regarding the pipeline route and related actions by the Nebraska legislature, the State Department announced a delay until 2013 of a national interest determination to gather additional information needed to assess a new pipeline route avoiding the Sand Hills. The Temporary Payroll Tax Cut Continuation Act of 2011(P.L. 112-78), enacted on December 23, 2011, included provisions requiring the Secretary of State to issue a permit for the project within 60 days, unless the President determined the project not to be in the national interest. The act allowed for future changes to the Nebraska route if approved by the Governor of Nebraska. On January 18, 2012, the State Department, with the President's consent, denied the Keystone XL permit, citing insufficient time under the 60-day deadline to obtain all the necessary information to assess the reconfigured project. TransCanada has stated that it will reapply for a Presidential Permit after a new proposed route through Nebraska is determined. In keeping with an agreement reached between TransCanada and Nebraska before the State Department's announcement, the company expects to establish the new route by October 2012. If the permit application process starts anew, a new draft EIS potentially could build upon the August 2011 final EIS, incorporating necessary analysis associated with a new Nebraska route.

Several legislative proposals have sought to influence or alter the existing permit process for Keystone XL. Several bills, like P.L. 112-78, would have required the President to issue a final order granting or denying the Presidential Permit for the Keystone XL pipeline by a specific deadline (S. 1932, H.R. 3400, H.R. 3537, and H.R. 3630). These provisions have been mooted by the State Department's denial of the permit. The North American Energy Access Act (H.R. 3548) would transfer permitting authority over the Keystone XL pipeline project from the State Department to the Federal Energy Regulatory Commission, and would require the commission to issue a permit for the project within 30 days of enactment. The Keystone For a Secure Tomorrow Act (H.R. 3811) would immediately approve the initial permit application filed by TransCanada in 2008. Both bills include provisions allowing for later alteration of the pipeline route in Nebraska. Both bills may also raise questions about the President's executive authority to issue cross-border permits.

INTRODUCTION[1]

In September 2008, TransCanada (a Canadian company) applied to the U.S. Department of State for a permit to cross the U.S.-Canada international border with the Keystone XL pipeline project. If constructed, the pipeline would carry crude oil produced from the oil sands region of Alberta, Canada, to U.S. Gulf Coast refineries. Because the pipeline would connect the United States with a foreign country, it requires a Presidential Permit issued by the State Department. Issuance of a Presidential Permit requires a finding that the project would serve the "national interest."

In the course of gathering information necessary to make its national interest determination, the State Department identified various concerns raised by the public. On November 10, 2011, the State Department announced its decision to seek additional information about alternative pipeline routes before it could move forward with a national interest determination.[2] More specifically, concerns regarding potential environmental impacts of constructing and operating the pipeline along the proposed route through the Sand Hills region of Nebraska led the State Department to decide that an assessment of potential alternative routes that would avoid that area was necessary. Subsequently, on November 14, 2011, TransCanada announced an agreement with the Nebraska Department of Environmental Quality to identify a pipeline route that would avoid the Sand Hills. The State Department estimated at the time that the preparation of supplemental environmental analysis necessary for a new route alternative could be complete in early 2013.

Subsequent to the State Department's announcement of a delay in the permit review, Congress acted to expedite a permit decision on the Keystone XL project. The Temporary Payroll Tax Cut Continuation Act of 2011(P.L. 112-78), enacted on December 23, 2011, included provisions requiring the Secretary of State to issue a permit for the project within 60 days, unless the President publicly determined the project not to be in the national interest. The act allowed for future changes to the Nebraska route if approved by the Governor of Nebraska. On January 18, 2012, the State Department, with the President's consent, denied the Keystone XL permit, citing insufficient time under the 60-day deadline to obtain all the necessary information to assess the reconfigured project.[3] TransCanada has announced that it will reapply for a Presidential Permit after a new proposed route through Nebraska is determined. The company expects to establish the new route by October 2012.[4]

Members of Congress have expressed support for the proposed pipeline's potential energy security and economic benefits while others have expressed reservations about its potential environmental impacts.[5] Though Congress, to date, has had no direct role in permitting the pipeline's construction, it may have an oversight role stemming from federal environmental statutes that govern the pipeline's application review process. Congress also may seek to influence the State Department permitting process, or may seek to assert direct congressional authority over permit approval, through new legislation.

A number of legislative proposals, like P.L. 112-78, would have imposed deadlines on a national interest determination for the Keystone XL project. The North American-Made Energy Security Act (H.R. 1938), would have directed the President to issue a final order granting or denying the Presidential Permit for the Keystone XL pipeline by November 1, 2011. The Jobs Through Growth Act (H.R. 3400), would have required the President to issue a final order granting or denying the Presidential Permit for the Keystone XL pipeline within 30 days of enactment. The North American Energy Security Act (S. 1932), which was introduced on November 30, 2011, would have required the Secretary of State to issue a permit for the project within 60 days of enactment, unless the President publicly determined the project to be not in the national interest. The North American Energy Security Act (H.R. 3537), introduced on December 1, 2011, and the Middle Class Tax Relief and Job Creation Act of 2011 (H.R. 3630), introduced on December 9, 2011, contain similar provisions for issuing a Presidential Permit within 60 days of enactment. All of these proposed provisions have been mooted by the State Department's denial of the permit.

The North American Energy Access Act (H.R. 3548), introduced on December 2, 2011, would transfer the permitting authority over the Keystone XL pipeline project from the State Department to the Federal Energy Regulatory Commission (FERC), requiring the commission to issue a permit for the project within 30 days of enactment. The Keystone For a Secure Tomorrow Act (H.R. 3811), introduced on January 24, 2011, would immediately approve the original permit application filed by TransCanada in 2008. Both bills include provisions allowing for later alteration of the pipeline route in Nebraska.

This report describes the Keystone XL pipeline proposal and the process required for federal approval. It summarizes key arguments for and against the pipeline put forth by the pipeline's developers, federal agencies, environmental groups, and other stakeholders. Finally, the report reviews the constitutional basis for the State Department's authority to issue a Presidential Permit, and opponents' possible challenges to this authority.

Pipeline Description

The U.S. portion of the Keystone XL pipeline project, as originally proposed, would pass through Montana, South Dakota, Nebraska, Oklahoma, and Texas (**Figure 1**). This route would consist of approximately 1,380 miles of 36-inch-diameter pipe and have the capacity to transport 830,000 barrels per day (bpd) of crude oil to the United States, delivering up to roughly 200,000 bpd to an existing oil terminal in Oklahoma with the remainder sent further to delivery points in Texas.[6] On November 14, 2011, TransCanada announced an agreement with the state of Nebraska to make as yet undetermined changes to the pipeline route in Nebraska to avoid environmentally sensitive areas.[7] These route changes are expected to increase the pipeline mileage through the state, although the company expects to maintain the original route through the other states, including its planned delivery points, so the pipeline's overall capacity should not be affected.

As of February 2011, the Keystone XL project along its original route was estimated to cost more than $7.0 billion, with the U.S. portion accounting for at least $5.4 billion of that total.[8] That is higher than the cost estimate when the initial permit application was filed reportedly due to currency swings, changing regulatory requirements, and permitting delays.[9] A new route would presumably be longer and cost more. The Keystone XL pipeline would be an extension of TransCanada's existing Keystone pipeline, which links the Alberta oil sands to refineries in Illinois and Oklahoma (**Figure 1**). The Keystone pipeline received State Department approval on March 17, 2008, and began commercial operation in June 2010.

Keystone XL Extension to Bakken Oil Production

The U.S. portion of the Bakken formation is an unconventional oil resource that underlies parts of North Dakota and Montana.[10] By the end of 2010, U.S. Bakken production was 350,000 bpd.[11] Output has climbed further in 2011. The oil is transported to refineries by rail and truck, rather than by pipeline, which would be more economic. In part, this is because infrastructure has not kept up with rapid production growth in the Bakken region in recent years. Output is expected to increase significantly in the future, increasing the need for pipeline transportation capacity.[12]

TransCanada has signed contracts with Bakken oil producers to carry 65,000 bpd from the region via the Keystone XL pipeline. While not the full 100,000 bpd of capacity TransCanada had offered to oil producers, this was enough to justify adding the Bakken Marketlink Project, a pipeline running from Baker, MT, to the Keystone XL pipeline, which can then carry crude to the oil hub at Cushing, OK, and on to the Gulf Coast.[13] The Bakken Marketlink would have a 100,000 bpd capacity and is estimated to cost $140 million. It could start operating in 2013 if it and the Keystone XL pipeline receive regulatory approvals.[14]

These new Bakken contracts also improve the economics for Keystone XL, raising the amount of oil slated to flow through the pipeline.[15] Lower transportation costs and access to new markets may support investment in the Bakken. Furthermore, TransCanada is not the only company adding pipeline capacity in the region. Notably, Enbridge, another Canadian pipeline company, has proposed the Bakken Pipeline Project, which will add 120,000 bpd of transport capacity to move Bakken oil to Midwest markets.[16] According to Enbridge, sufficient pipeline capacity has been slow to emerge in the region because "they're smaller players in the Bakken. They are no able to make the 20-year commitments and it's been a lot of work to get them to commit to the level that [is] required to underwrite a major project out of the Bakken."[17] Rail transport capacity is also expanding.[18]

Presidential Permit Application Requirements and Status

Ordinarily, the U.S. government does not have permit authority for oil pipelines, even interstate pipelines. This is in contrast to interstate natural gas pipelines, which, under Section 7(c) of the Natural Gas Act, must obtain a "certificate of public convenience and necessity" from the Federal Energy Regulatory Commission.[19] Generally, the primary siting authority for oil pipelines would be established under applicable state law (which may vary considerably from state to state). However, the construction, connection, operation, and maintenance of a pipeline that connects the United States with a foreign country requires executive permission conveyed through a Presidential Permit. Since the Keystone and proposed Keystone XL pipelines are designed for the importation of oil from Canada, their facilities require a Presidential Permit.

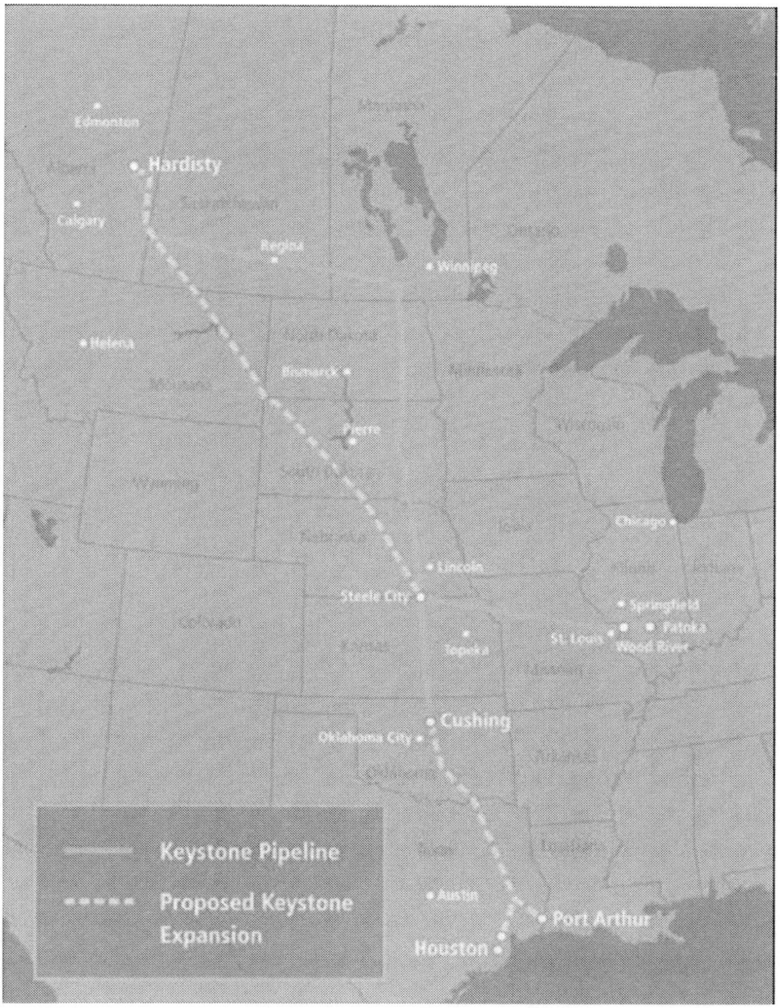

Source: U.S. Department of State, http://keystonepipeline-xl.state.gov/clientsite/keystonexl.nsf/map.jpg? OpenFileResource.

Note: Figure 1 shows the developer's originally proposed "preferred alternative" for the Keystone XL pipeline route according to Presidential Permit application documents, however, the route through Nebraska is expected to change. For discussion of alternative routes, see the State Department EIS discussed below.

Figure 1. TransCanada Keystone Pipeline and Original Keystone XL Proposed Route.

Executive Order 13337 delegates to the Secretary of State the President's authority to receive applications for Presidential Permits.[20] Issuance of a Presidential Permit is dependant upon a finding that the project would serve the national interest.[21] In the course of making that determination, the State Department is obligated to consider a host of issues related to the proposed project including its potential impacts to the environment, economy, energy security, and foreign policy, to name a few. In that capacity, the State Department is required to consult with relevant federal and state agencies and to invite public comment in arriving at its determination. Ultimately, however, the State Department has discretion in determining what factors to examine to inform its determination of whether a proposed project is in the national interest (i.e., will be granted a Presidential Permit).

Identifying and considering the potential environmental impacts of a proposed project is donewithin the context of the State Department's preparation of an Environmental Impact Statement (EIS), as required pursuant to the National Environmental Policy Act (NEPA, 42 U.S.C. §4321 et seq.).[22] After issuing a final EIS, a public review period begins during which the State Department receives comments from the public and local, state, tribal, and federal agencies. As it is implemented by the State Department, that public comment process is part of the process of making its national interest determination, not a requirement under NEPA.

On November 11, 2011, in response to public comments on the final route of the Keystone XL pipeline, the State Department announced its decision that certain environmental issues identified in the final EIS and further stressed in public comments led it to seek additional information about an alternative pipeline route before making its final national interest determination.[23]

Background regarding NEPA requirements applicable to the State Department's obligation to identify and assess the potential environmental impacts of the proposed Keystone XL pipeline, as well as the effect that the identification of those impacts has had the process to make a national interest determination, are discussed below.

Background and Overview of the NEPA Process for the Pipeline Project

Broadly, NEPA requires federal agencies to consider the environmental impacts of their actions before proceeding with them and to inform the public of those potential impacts. To ensure that environmental impacts are considered, an EIS must be prepared for major federal actions "significantly" affecting the environment.[24] With respect to the application submitted by

TransCanada, the State Department concluded that issuance of a Presidential Permit for the proposed construction, connection, operation, and maintenance of the Keystone XL Pipeline and its associated facilities at the United States border would constitute a major federal action that may have a significant impact upon the environment within the meaning of NEPA.[25] For this reason, the State Department prepared an EIS to address reasonably foreseeable impacts from the proposed action and alternatives.

Among other requirements, an EIS must include a statement of the purpose and need for an action, a description of all reasonable alternatives to meet that purpose and need, a description of the environment to be affected by those alternatives, and an analysis of the direct and indirect effects of the alternatives, including cumulative impacts.[26] Accordingly, the State Department EIS must review and consider the potential environmental impacts of the entire pipeline (including the construction, operation, and maintenance of the pipeline and its associated facilities), not just the facilities at the border crossing.

As the NEPA compliance process for TransCanada's permit application has proceeded, it is important to understand the distinction between what is required under NEPA itself and what may be required pursuant to other environmental requirements identified within the context of the NEPA process. NEPA itself requires federal agencies to identify the environmental impacts of an action before proceeding with them and to involve the public in that process when environmental impacts are significant. In that process of identifying a proposed project's environmental impacts, as evidenced in environmental reviews, studies, or analyses included in the EIS, compliance obligations established under additional local, state, tribal and federal law would also be identified.

The Draft Environmental Impact Statement

NEPA regulations require preparation of a draft EIS that must be circulated for public and agency comment, followed by a final EIS that incorporates those comments.[27] Preparing the EIS is the responsibility of a designated "lead agency," in this case, the State Department. In developing the EIS, the State Department must rely to some extent on information provided by TransCanada. For example, TransCanada's permit application included an Environmental Report which was intended to provide the State Department with sufficient information to understand the scope of potential environmental impacts of the project.[28]

The EIS must also identify any state, tribal, or federal licenses, permits or approvals applicable to the project in the United States.[29] Further, in preparing the draft EIS, the lead agency must request input from "cooperating agencies" which include any agency with jurisdiction by law or with special expertise regarding any environmental impact associated with the project.[30] Cooperating agencies for the Keystone XL project are the U.S. Environmental Protection Agency (EPA); the Department of Transportation's Pipeline and Hazardous Materials Safety Administration (PHMSA), Office of Pipeline Safety (OPS); the Department of the Interior's Bureau of Land Management, U.S. Fish and Wildlife Service, and National Park Service; the U.S. Army Corps of Engineers; the U.S. Department of Agriculture's Farm Service Agency, Natural Resources Conservation Service, and Rural Utilities Service; the Department of Energy's Western Area Power Administration; and state environmental agencies.

In addition to its role as a cooperating agency, EPA is also required to review and comment publicly on the EIS and rate both the adequacy of the EIS itself and the level of environmental impact of the proposed project.[31] Rating the EIS takes place after the draft is issued. The EIS could be rated either "Adequate," "Insufficient Information," or "Inadequate." EPA's rating of a project's environmental impacts may range from "Lack of Objections" to "Environmentally Unsatisfactory" (EPA rating of environmental impacts is discussed in more detail, below).

The State Department released its draft EIS for the proposed Keystone XL Pipeline project for public comment on April 16, 2010.[32] The draft EIS identified TransCanada's "preferred alternative" (**Figure 1**) for the project as well as other alternatives considered. On July 16, 2010, EPA rated the draft EIS "Inadequate."[33] EPA found that potentially significant impacts were not evaluated and that the additional information and analysis needed was of such importance that the draft EIS would need to be formally revised and again made available for public review. Additional criticism of the State Department's implementation of the NEPA process followed an October 21, 2010, statement by Secretary Clinton that, while analysis of the project was not complete and a final decision had not been made, the State Department was "inclined to" approve the project.[34] Critics of the project, including some Members of Congress, stated that the Secretary's statement appeared to prejudge its permit approval for the pipeline proposal as a foregone conclusion.[35]

The Supplemental Draft Environmental Impact Statement

The State Department issued a supplemental draft EIS on April 15, 2011. In addition to addressing issues associated with EPA's inadequacy rating, the supplemental draft EIS addressed comments received from other agencies and the public. On June 6, 2011, EPA sent a letter to the State Department that rated the supplemental draft EIS as having "Environmental Objections—Insufficient Information."36 EPA acknowledged that the State Department had "worked diligently" to develop additional information in response to EPA's comments and the large number of other comments on the draft EIS. However, EPA believed that additional analysis needed to be included in the final EIS to fully respond to its earlier comments. Among other items, EPA recommended that the State Department should do the following: improve the analysis of the potential oil spill risks, including additional analysis of other reasonable alternatives to the proposed pipeline route; provide additional analysis of potential oil spill impacts, health impacts, and environmental justice concerns to communities along the pipeline route and adjacent refineries; and improve its characterization of lifecycle greenhouse gas emissions associated with Canadian oil sands crude.

In its June 6th letter to the State Department, EPA refers to agreements with the State Department that certain deficiencies identified in the supplemental draft EIS would be addressed in the final EIS. Further, in its conclusion, EPA stated that it would carefully review the final EIS to determine if it fully reflects those agreements and if measures to mitigate adverse environmental impacts are fully evaluated.

The Final Environmental Impact Statement

On August 26, 2011, the State Department issued the final EIS for the proposed Keystone XL Pipeline. Among other elements of the final EIS, it identified various major pipeline route alternatives and an environmental analyses of potential impacts associated with those alternatives.[37]

Generally, after a federal agency issues a final EIS, EPA would be required to rate the environmental impacts of the project based on the findings in that EIS. The project may receive a rating of either "Lack of Objections," "Environmental Concerns," "Environmental Objections," or "Environmentally Unsatisfactory." The federal agency would then be required to respond to EPA's rating, as appropriate. EPA did not rate or provide comments on the final EIS before the decision was made to change the pipeline's proposed route through Nebraska. Given the State Department's subsequent denial of the

Keystone XL permit, EPA may not submit formal to the State Department on the final EIS.

The Role of Environmental Impacts on the National Interest Determination

Following the release of the Keystone XL project's final EIS, a review period began to determine if the proposed project was in the national interest. According to the State Department, the period involves a broader evaluation of the permit application that extends beyond the project's environmental impacts, taking into account economic, energy security, foreign policy, and other relevant issues. That is, it takes into account the range of issues the State Department deems necessary to determine if the proposed project is in the national interest. As part of this process for the Keystone XL project, the State Department held public meetings in each of the six states through which the proposed pipeline would pass and in Washington, DC.[38] The meetings were intended to give members of the public additional opportunity to voice their opinions on issues they thought should be taken into account in determining whether granting or denying the Presidential Permit would be in the national interest.[39] During the review period, the State Department received input from state, local, and tribal officials as well as members of the public.

Generally, after a final EIS is issued, a final project decision would be reflected in a "Record of Decision and National Interest Determination," issued by the State Department. A Record of Decision (ROD) is issued pursuant to NEPA. It formalizes the selection of a project alternative. For Presidential Permit applications for pipeline projects, a ROD and national interest determination are usually issued as the same document.[40] However, after the public review period, the State Department issued a statement regarding the public comments and its response to those comments.[41] During the public review period, the State Department stated that it received comments on a wide range of issues including the Keystone XL project's potential impact on jobs, pipeline safety, health concerns, the societal impact of the project, and oil extraction in Canada. Concern regarding the proposed pipeline route through the Sand Hills area of Nebraska was identified as one of the most common issues raised. Comments regarding that pipeline route were consistent with the environmental impacts identified in the final EIS with regard to the unique combination of characteristics of the Sand Hills region (e.g., a high concentration of wetlands of special concern, a sensitive ecosystem, and extensive areas of very shallow groundwater, discussed in more detail in the "Impacts to the Nebraska Sand Hills" section, below). Further, the Nebraska legislature was convening a special session to consider the legislation that would establish regulations applicable to pipeline siting within the state. Facing the prospect of new state pipeline siting regulations applicable to the Sand Hills, together with the concern about the Keystone XL pipeline's specific "preferred" route, the State Department determined it would be necessary to examine alternative routes that would avoid this region in Nebraska.

The State Department's decision to gather additional information regarding the alternative pipeline routes, after the final EIS was issued, illustrates the distinctly different, yet interrelated requirements applicable to the publication of a ROD and the national interest determination. Under NEPA, the State Department must fully assess the environmental consequences of an action and potential project alternatives *before* making a final decision. NEPA does not prohibit a federal action that has adverse environment impacts; it requires only that a federal agency be fully *aware of* and *consider* those adverse impacts before selecting a final project alternative. That is, NEPA is intended to be part of the decision-making process, not dictate a particular outcome. The State Department's decision to issue a Presidential Permit, however, dictates a particular outcome—that a Permit will not be granted unless it is determined that the project is in the national interest. While NEPA does not prohibit federal actions with adverse environmental impacts, a project's adverse environmental impacts (as well as other factors) may lead the State Department to determine that it is not in the national interest.

The NEPA Process for a New Permit Application

In the wake of the State Department's denial of TransCanada's Presidential Permit application (discussed below), it is anticipated that TransCanada will submit a new application for a Presidential Permit. If that occurs, the reapplication process would require a "new" NEPA process (i.e, preparation of a draft and final EIS). However, much of the analysis included in the August 26th final EIS could potentially form the bulk of the draft EIS, with the inclusion of analysis of environmental impacts associated with new pipeline routes through Nebraska or any additional information that may be relevant to a new proposal. The State Department, however, has declined to state that a new Keystone XL permit review would be expedited on this basis, although the department has stated that it does have guidelines to use information already available in a permit review.[42] Nebraska officials reportedly have stated that they could decide on a new pipeline route through the state by September 2012 at the earliest.[43]

Since EPA did not comment on the final EIS, it is unclear if the issues raised regarding the supplemental EIS were addressed to EPA's satisfaction. However, if TransCanada reapplies for a pipeline permit and if EPA did have comments or concerns regarding the August 26th final EIS (apart from issues related to the previous pipeline route through Nebraska), early identification of those issues would be useful in expediting the NEPA process for the reapplication.

Delays and Deadlines in Keystone XL Permit Review

Shortly after issuing its final EIS, the State Department indicated that it expected to reach a final decision on whether to grant the Keystone XL permit before the end of 2011. The North American-Made Energy Security Act (H.R. 1938), which passed the House on July 7, 2011, would have directed the President to expedite the State Department's permit review process, requiring a final decision to grant or deny the permit no later than November 1, 2011 (§3(c)). H.R. 1938 was motivated by the perception among some in Congress that the State Department was taking too long to review an energy infrastructure project critical to national security and economic growth.[44] Opponents of the bill argued that the project's unique and potentially unacceptable safety and environmental risks, as well as its uncertain impacts on fuel prices, required more time for analysis and evaluation.[45] The bill was not voted upon in the Senate, however, prior to the proposed November 1 deadline.

The Temporary Payroll Tax Cut Continuation Act of 2011(P.L. 112-78), enacted on December 23, 2011, included provisions requiring the Secretary of State to issue a permit for the project within 60 days, unless the President publicly determined the project to be not in the national interest. Other legislative proposals, which were not enacted, similarly sought to expedite the permit decision. The Jobs Through Growth Act (H.R. 3400), which was introduced on November 10, 2011, would have required the President to issue a final order granting or denying the Presidential Permit for the Keystone XL pipeline within 30 days of enactment (§396(b)). Three subsequent bills, S. 1932, H.R. 3537, and H.R. 3630, would all have mandated that the Secretary of State issue a Keystone XL permit within 60 days of enactment, unless the President publicly determined the project to be not in the national interest. The three bills would have deemed the current final EIS adequate (concluding the federal environmental review process) but would have required the permit to recognize an alternative pipeline route approved in the future by Nebraska, while not delaying construction elsewhere. Because the State Department has already reached its decision regarding the Keystone XL project permit, these deadline proposals have become moot.

In October 2011, fourteen Members of Congress wrote to the State Department's Office of Inspector General requesting an investigation of the department's handling of the EIS and National Interest Determination for the Keystone XL project.[46] The request was prompted, in part, by press reports suggesting bias or potential conflicts of interest in the State Department's hiring of an outside contractor to perform the EIS and in its communications with the pipeline's developer, TransCanada.[47] On November 4, the Inspector General's Office announced that, in response to this request, it was initiating a special review "to determine to what extent the Department and all other parties involved complied with Federal laws and regulations relating to the Keystone XL pipeline permit process."[48] The announcement did not indicate, however, how long this review would take and what impact it might have on the Presidential Permit application review schedule.

Denial of the Keystone XL Permit

On November 10, 2011, the State Department announced that it would require additional information about alternative pipeline routes avoiding the environmentally sensitive Sand Hills area in Nebraska before moving forward with its National Interest Determination.[49] Although the State Department did not decide that environmental issues led to a determination that the proposed project was not in the national interest, environmental issues identified in the final EIS, and further stressed in public comments, led to its decision to delay that determination until it gathered this information. In a concurrent press release, President Obama stated

> Because this permit decision could affect the health and safety of the American people as well as the environment, and because a number of concerns have been raised through a public process, we should take the time to ensure that all questions are properly addressed and all the potential impacts are properly understood.[50]

Although no new decision deadline was established, State Department officials suggested that it would be "reasonable to expect that this process including a public comment period on a supplement to the final EIS consistent with NEPA could be completed as early as the first quarter of 2013."[51] In a prior press interview, President Obama also appeared to suggest that, notwithstanding the delegation of Presidential Permit authority to the State Department, he would be personally involved in the final decision on the Keystone XL Pipeline permit application.[52]

As noted earlier, the State Department, with the President's consent, denied the Keystone XL permit on January 18, 2012. In it's announcement the department stated that its decision "was predicated on the fact that [P.L. 122-78] ... passed in December does not provide sufficient time to obtain the information that we think is necessary to assess whether the project, in its current state, is in the national interest."[53] However, the department also stated that it's decision did not preclude TransCanada from reapplying for a Presidential Permit in the future, although such a reapplication "will trigger... a completely new review process."[54]

In light of the State Department's denial of the Keystone XL permit, some in Congress seek alternative means to support development of the pipeline. The North American Energy Access Act (H.R. 3548), rather than imposing a deadline on the State Department, would transfer the permitting authority over the Keystone XL pipeline project from the State Department to the Federal Energy Regulatory Commission (FERC), requiring the commission to issue a permit for the project within 30 days of enactment. Changing the State Department's role in issuing crossborder infrastructure permits may raise questions about the President's executive authority, however (further discussed in the **Appendix**).

In response to H.R. 3548, the State Department's key official on Keystone XL testified before Congress that

> The legislation raises serious questions about existing legal authorities, questions the continuing force of much of the federal and all of the state and local environmental and land use management authority over the pipeline, and overrides foreign policy and national security considerations implicated by a cross border permit, which are properly assessed by the State Department.[55]

The proposal may also raise some administrative and legal challenges for FERC. A senior commission official testified that the proposed legislation does not provide enough time for an "adequate" public record, provides no clear authority for enforcing measures required in the EIS, does not articulate a process for authorizing alterations to the pipeline route in Nebraska, and is unclear about permits required from other federal agencies, among other concerns.[56]

Given the State Department's permit denial, legal challenges are a possibility. However, in the event of a challenge based on an environmental issue, the distinction between State Department actions required under NEPA and those required under its authority to issue a Presidential Permit would be relevant. NEPA does not create a private right of action. Instead, judicial challenges to a federal agency action under NEPA are brought pursuant to the Administrative Procedure Act (APA, 5 U.S.C. §§706 et seq.). Presidential actions, however, are not subject to judicial review under the APA.[57] So the final EIS associated with the Keystone XL Pipeline may be subject to judicial review, but the State Department's national interest determination, made under its authority to issue a Presidential Permit, is not. For more information regarding the State Department's authority to grant a Presidential Permit, see the **Appendix.**

State Siting and Environmental Approvals

As noted above, the federal government does not currently exercise siting authority over oil pipelines. Siting for the Keystone XL pipeline still must comply with any applicable state laws. These laws vary from state to state. South Dakota, for example, required TransCanada to apply for a permit for the Keystone XL pipeline from the state public utility commission, which issued the permit on April 25, 2010.[58] Montana requires a certificate from the state's Department of Environmental Quality.[59] At the time of TransCanada's application for a Presidential Permit, Nebraska did not have any permitting requirements that applied specifically to the construction and operation of oil pipelines, although a state statute does include an "eminent domain" provision, which grants eminent domain authority to oil pipeline companies that are unable to obtain the necessary property rights from the relevant property owners.[60] Due to the controversy surrounding the Keystone XL project, Nebraska called a special session of its legislature to enact legislation to assert state authority over pipeline siting.[61] A number of additional approvals and permits required by the states along the proposed route are summarized in TransCanada's Presidential Permit application.[62]

ARGUMENTS FOR AND AGAINST THE PIPELINE

Proponents of the Keystone XL pipeline, including Canadian agencies and U.S. and Canadian petroleum industry stakeholders, base their arguments supporting the pipeline primarily on increasing the diversity of the U.S. petroleum supply and economic benefits to the United States, including job creation. Environmental groups object to the project on the grounds that Canadian oil sands development has negative environmental impacts—particularly that development of oil sands releases more greenhouse gases than development of alternatives, like conventional oil or renewable fuels (discussed in greater detail below). Also, they argue that it promotes continued U.S. dependency on fossil fuels and affected communities along the route. Some members of the public are opposed to the selection of the proposed pipeline route. Opposition generally stems from concerns regarding potential impacts associated with the pipeline's construction and operation in communities along the proposed route—particularly potential impacts to land use (e.g., cattle grazing) and groundwater contamination in the event of a release. As mentioned above, concern that ultimately delayed the State Department's decision to make a national interest determination stemmed from concern regarding a segment of the route that crossed the Sand Hills region of Nebraska.

Impacts to the Nebraska Sand Hills

In the process of examining factors necessary to determine whether the Presidential Permit for the Keystone XL pipeline is in the national interest, the State Department decided that it needs to assess potential alternative pipeline routes that would avoid the Sand Hills region of Nebraska. Unique characteristics of the Sand Hills, including its high concentration of wetlands, extensive areas of very shallow groundwater, and its sensitive ecosystem, were identified as factors that resulted in increasing public concern over the proposed pipeline location. Subsequently, TransCanada announced that it would work with the Nebraska Department of Environmental Quality to identify a potential pipeline route that would avoid the Nebraska Sand Hills.[63]

To understand the concerns associated with potential environmental impacts of the construction and operation of a pipeline that crosses the Sand Hills (also referred to as the Sandhills) an understanding of the unique size and structure of the region is useful. The Sand Hills region is a 19,600 square mile sand dune formation stabilized by native grasslands that cover 95% of its surface. The surface is highly susceptible to wind erosion if the grassland is disturbed.[64] Below its surface lie hundreds of feet of course sand and gravel. Essentially, the porous soil acts like a giant sponge that quickly absorbs precipitation, allowing very little to run off. In some areas, the water table reaches the land surface—a characteristic that creates lakes that dot the region as well as 1.3 million acres of wetlands.

The loose, porous soil and sensitivity to wind erosion have been factors contributing to a lack of development on the Sand Hills. As a result, the region contains the most intact natural habitat of the Great Plains of the United States. The porosity of the soil is also relevant because the Sand Hills sits atop the Ogallala Aquifer—one of the largest aquifer systems in the world.[65] In the final EIS, the preferred pipeline route through Nebraska would be located entirely above the Ogallala Aquifer (**Figure 2**).

The highly porous soil of the Sand Hills make it a significant recharge zone in the northern region of the Ogallala Aquifer system. That is, the sandy, porous soil of the Sand Hills allows a significant amount of surface water to enter (recharge) the aquifer system. Water from the aquifer also accounts for a significant amount of water use—78% of the region's public water, 83% of irrigation water in Nebraska, and 30% of water used in the U.S. for irrigation and agriculture.

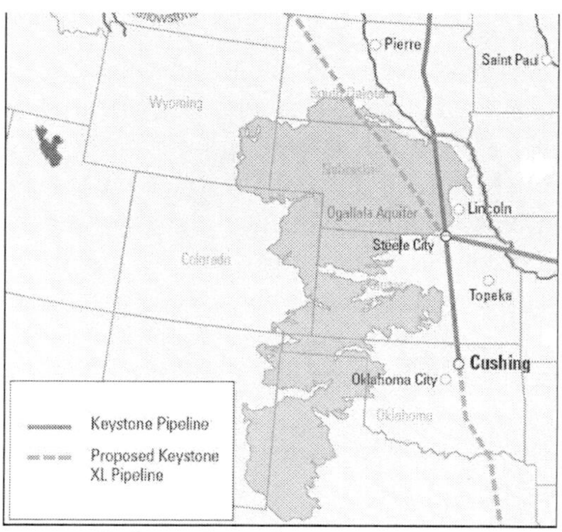

Source: Natural Resources Defense Council, *Say No to Tar Sands Pipeline*, November, 2010, p. 3.

Figure 2. Keystone XL Pipeline Route Across the Ogallala Aquifer.

Potential impacts to the Ogallala Aquifer and the Sand Hills identified in the final EIS include potential groundwater contamination after a release (e.g., a spill or leak from a hole or damaged portion of the pipeline) of crude oil during the construction or operation of the proposed pipeline. Along the preferred route of the proposed pipeline, areas in the Sand Hills region were identified as locations where the water table may be close to the surface. The depth to groundwater is less than 10 feet for approximately 65 miles of the preferred pipeline route in Nebraska. Both the soil porosity and the close proximity of groundwater to the surface increase the potential that a release of oil from the pipeline could contaminate groundwater in the region.[66]

Impact on U.S. Energy Security

In its Presidential Permit application, TransCanada asserts that constructing the proposed Keystone XL pipeline is in the U.S national interest to maintain adequate crude oil supplies for U.S. refineries. The application argues that the pipeline will allow U.S. refiners to substitute Canadian supply for other foreign crude supply and to obtain direct pipeli access to secure and growing Canadian crude output. In particular, the application asserts that the pipeline would allow the United States to decrease its dependence on foreign crude oil supplies from Mexico and Venezuela, the two largest oil importers into the U.S. Gulf Coast.[67]

These arguments have taken on additional weight in light of the recent political unrest in the Middle East and North Africa. However, it is worth noting that even if Keystone XL is built, prices for the crude oil it carries as well as domestically produced oil from elsewhere will continue to be affected by international events. The oil market is globally integrated and events in major producer and consumer countries can affect prices everywhere.[68] For example, the disruption of Libyan supply in early 2011 contributed to higher crude oil prices in the United States, even though the United States imported almost no oil from Libya before the unrest broke out.[69]

Canadian Oil Imports in the Overall U.S. Supply Context[70]

Gross U.S. imports of crude oil and petroleum products averaged 11.8 million barrels per day (Mbpd) in 2010.[71] Exports averaged 2.3 Mbpd, leaving net imports at 9.4 Mbpd.[72] U.S. net imports declined each year between 2005 and 2010 as a result of lower total oil demand and higher domestic supply. Domestic demand has decreased by about 1.7 Mbpd versus 2005 levels due largely to the economic recession. Meanwhile, U.S. production of oil and oil alternatives (including crude oil, natural gas liquids, and biofuels) increased by 1.4 Mbpd between 2005 and 2010. As a result, net imports fell by roughly 3.1 Mbpd since 2005.[73] Some of this decline could be mitigated in the near term as oil demand recovers from the recession or if domestic supply were to fall. However, there is increasing consensus among forecasters that U.S. net oil imports have passed their high water mark already and may remain relatively flat in the long run.[74]

Among the largest sources of U.S. gross oil imports are Canada (2.5 Mbpd), the Persian Gulf (1.7 Mbpd), and Mexico (1.3 Mbpd). Imports from the latter two sources have decreased in recent years in part due to lower need for imports described above and in part due to developments in those countries (**Figure 3**). All major Persian Gulf exporters are members of the Organization of the Petroleum Exporting Countries (OPEC), which cut production in 2009 to support oil prices. Mexican production has been falling since 2004 because new oil developments have not been able to offset depletion at Mexico's giant Cantarell field. Imports from Venezuela, another key source of U.S. imports, have also fallen. Venezuelan production never fully recovered after a strike at its national oil company, *Petróleos de Venezuela*, in 2002-2003. Venezuelan production today is nearly 1 Mbpd less than that achieved in 2001. In recent years, Venezuela has also been trying to diversify business away from the United States, for example, by increasing exports to China.[75]

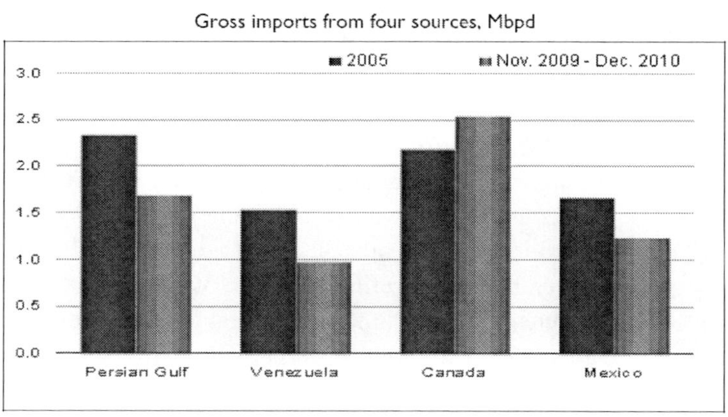

Source: U.S. Energy Information Administration, "*Petroleum Navigator: U.S. Imports by Country of Origin*," December 12, 2010, http://www.eia.gov/dnav/pet/pet_move_impcus_a2_nus_ep00_im0_mbblpd_m.htm.

Figure 3. U.S. Changes in U.S. Oil Imports, Selected Sources.

Monthly imports in Mbpd on a 12-month moving average, Jan. 2000 to Oct. 2010

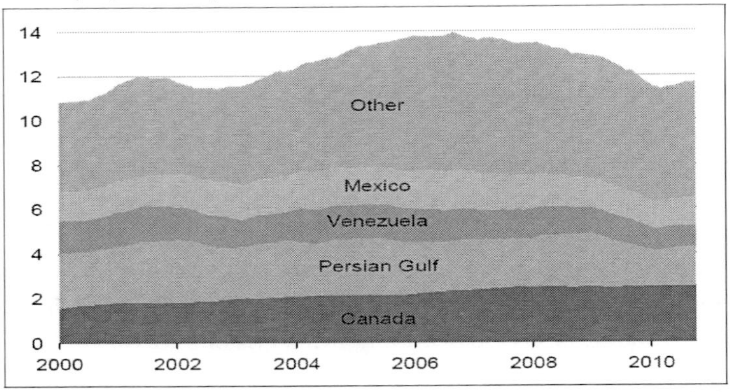

Source: U.S. Energy Information Administration, "*Petroleum Navigator: U.S. Imports by Country of Origin*," December 12, 2010. http://www.eia.gov/dnav/pet/pet_move_impcus_a2_nus_ep00_im0_mbblpd_m.htm.

Figure 4. Gross U.S. Oil Imports.

Meanwhile, Canadian production and exports to the United States have increased, primarily due to growing output from the oil sands in western Canada. Energy markets in the United States and Canada are well integrated by pipeline infrastructure, and nearly all Canadian energy exports go to U.S. consumers.[76] Canadian oil production has increased about 0.2 Mbpd since 2005, and exports to the United States increased by 0.4 Mbpd (**Figure 4**).[77] Canadian oil production is expected to grow by as much as 1.6 Mbpd between 2009 and 2025, mostly through increased output from the oil sands.[78]

Oil Sands, Keystone XL, and the U.S. Crude Oil Market

Oil sands (also referred to as tar sands) are a mixture of clay, sand, water, and heavy black viscous oil known as bitumen. Oil sands require more processing than conventional crude oil. Oil sands are processed to extract the bitumen, which can then be sent to refine es in one of two forms. Bitumen can be upgraded into "syncrude," a light crude that is suitable for pipeline transport and is relatively easy to refine. Alternatively, bitumen can be blended with lighter hydrocarbons to form a heavy crude that can be transported by pipeline. The bulk of oil sands supply growth is expected to be in the form of the latter.[79]

Most oil sands imports into the United States currently go to the Midwest, where some refineries are investing in complex refining capacity to process growing volumes of heavy Canadian crude.[80] The U.S. Gulf Coast region already has a large amount of complex refining capacity and is considered potentially well suited for processing Canadian heavy crude oil.[81] Gulf Coast refiners currently process heavy crudes from Venezuela, Mexico, and elsewhere. Complex refineries in the Gulf Coast may be best equipped to handle a large increase of heavy oil sands crude, though they may still need to adjust processes and make new capital investments in equipment to accommodate particular crudes' characteristics,[82] especially if the new Canadian crudes will be used in large amounts.[83] There are 15 refineries within Keystone XL's proposed delivery area in Texas that currently process heavy crude oil which is similar in composition to the oil that Keystone XL pipeline would carry.[84]

Oil production from the oil sands is increasing as is production from the Bakken and other areas of the U.S. Midwest.[85] Transport options to ship crude from the Midwest to the Gulf Cost are limited. (Traditionally, crude oil had been shipped up from the Gulf Coast to Midwestern refineries). The resulting abundance of crude oil in the Midwest has driven down crude oil prices in that region relative to Gulf Coast and international crude prices. Midwestern refiners benefit from the lower cost of crude. This, however, does not translate to substantially lower gasoline or other products prices for consumers in the Midwest. The Midwest still brings in refined products from other regions which keeps refined products prices in line with national and international levels.[86]

Oil sands producers are interested in Keystone XL because it would expand their market reach into the Gulf Coast region. This would increase the number of refineries they could sell their crude to because Gulf Coast refineries have more upgrading capacity which is better suited to run heavy crude. Reaching a larger market and one with more upgrading capacity could increase the price these producers receive for their crude. Gulf Coast refiners are interested in the Keystone XL pipeline because it increases the supply of heavy sour crude in the Gulf region, potentially bringing down their input costs relative to the options they currently have available. Canadian Natural Resources Limited, an oil sands producer, and Valero Energy Corporation, a large U.S. refiner, are among those that contracted for shipping capacity on the Keystone XL pipeline.

With expanded pipeline capacity extending to the U.S. Gulf Coast, Canadian oil sands crude may compete with other heavy crudes such as those from Mexico, Venezuela, and elsewhere.[87] It is difficult to predict precisely how this competition will play out, but it may take place through shifting discounts or premiums on crude oils from various sources.[88] It may be possible for Canadian oil supplies to effectively "push out" waterborne shipments from other countries, although this depends on a wide range of market conditions. Waterborne crudes may more easily go to other destinations than Canadian crudes, though like Canadian crudes they can be tied to specialized refining capacity, as is true for Venezuelan heavy crudes.

There is concern that this increased supply of crude to the Gulf Coast may result in larger refined product exports rather than contributing to lower domestic fuel cost. Although the United States is a net importer of oil and oil products, it does export some refined products.[89] U.S. refined product exports rose when domestic demand declined in the wake of the recession and as foreign demand for certain fuels, like diesel, has remained relatively robust. If Keystone XL secures growing oil sands output to flow into the United States, it could push out seaborne crudes from elsewhere, regardless of where the product is ultimately sold. If the absence of the pipeline encourages Canadian oil sands producers and pipeline companies to find an alternate export route through the Canadian West Coast, Canadian supplies may displace heavy oil supplies in other markets and potentially allow relatively more overseas imports coming into the Gulf Coast. This possibility is discussed further below.

Other Pipeline Projects

Apart from Keystone XL, several other pipeline proposals could help carry growing Canadian crude oil supplies to the U.S. Gulf Coast. On October 16, 2011, Enbridge announced it would purchase ConocoPhillips share of the Seaway pipeline and reverse its direction to bring crude oil from the Midwest to the Gulf Coast. ConocoPhillips had kept the pipeline running northward to serve its refinery in Ponca City, OK. However, the glut of oil in the Midwest had resulted in the pipeline running at low volumes. Nonetheless, ConocoPhillips had been uninterested in reversing the pipeline. ConocoPhillips, which is spinning off its refining business,[90] is now selling its share of Seaway to Enbridge. Enbridge and Seaway shareholder Enterprise Products Partners L.P. immediately announced that they would reverse the direction of crude oil flows on the Seaway pipeline to enable it to transport oil from Cushing, OK, to the U.S. Gulf Coast. Pending regulatory approval, the line could operate in reversed service with an initial capacity of 150,000 barrels per day by second quarter 2012.[91]

Prior to the Seaway sale, Enbridge has reported it has significant commitments for two new pipeline projects, Flanagan South and Wrangler, that would carry oil from Illinois to Oklahoma and then from Oklahoma to

Texas.[92] According to Enbridge, the project would duplicate existing routes and does not cross an international border, in which case it does not require a Presidential Permit. Enbridge already has cross border pipeline capacity connecting Alberta to Illinois.

However, according to reports, Wrangler has been canceled in light of the Seaway purchase and reversal.[93] Like Keystone XL, these projects would also have facilitated increased flow of Canadian crude to the U.S. Gulf Coast.

Canadian Oil to Alternative Markets

In 2010, 98% of Canada's oil exports went to the United States, mostly through north-south pipelines. One major oil pipeline extends from Alberta to Canada's west coast: the Trans Mountain Pipeline, which is owned by Houston-based Kinder Morgan and has a capacity of 300,000 bpd. Some of the oil from the Trans Mountain Pipeline is loaded onto tankers and shipped from Vancouver. Currently, about 90% of the crude shipped out by sea goes to California, with the remainder shipped to the U.S. Gulf Coast and Asia.[94]

There are proposals to increase the capacity for oil from Alberta to reach the Canadian west coast. Kinder Morgan is considering expanding the Trans Mountain Pipeline to 700,000 bpd, more than doubling its existing capacity, and expanding west coast shipping facilities. Enbridge has proposed a new pipeline: the Northern Gateway project would have a 525,000 bpd capacity to send oil from Edmonton to Kimat, British Columbia.[95] These projects reflect anticipated growth of western Canadian oil production and an interest by Canadian oil producers to diversify their available markets beyond U.S. customers. Both proposals have received criticism from environmentalists. Because it would require construction of a completely new pipeline, Northern Gateway in particular has been criticized by some environmental and First Nations groups.[96]

If export capacity to the west coast is expanded it could increase the amount of Canadian crude oil going to non-U.S. markets. Canadian interests assert that Canadian oil sales to Asian markets, where oil demand is growing rapidly, are more likely if greater shipments to the United States are not possible.[97] A study commissioned by the U.S. Department of Energy suggested that

> if pipeline projects to the BC [British Columbia] coast are built, they are likely to be utilized. This is because of the relatively short marine distances to major northeast Asia markets, future expected growth there in refining capacity and increasing ownership interests by

> Chinese companies especially in oil sands production. Such increased capacity would alter global crude trade patterns. Western Canadian Sedimentary Basin (WCSB) crudes would be "lost" from the USA, going instead to Asia. There they would displace the world's balancing crude oils, Middle Eastern and African predominantly OPEC grades, which would in turn move to the USA. The net effect would be substantially higher U.S. dependency on crude oils from those sources versus scenarios where capacity to move WCSB crudes to Asia was limited.[98]

Economic Impact of the Pipeline

In addition to supply diversity arguments, some Keystone XL pipeline proponents support the project based on economic benefits associated with expanding U.S. pipeline infrastructure. A recent study by the Energy Policy Research Foundation, for example, concludes that "the Keystone expansion would provide net economic benefits from improved efficiencies in both the transportation and processing of crude oil of $100 million-$600 million annually, in addition to an immediate boost in construction employment."[99] A 2009 report from the Canadian Energy Research Institute (CERI) commissioned by the American Petroleum Institute similarly concludes that

> As investment and production in oil sands ramps up in Canada, the pace of economic activity quickens and demand for US goods and services increase rapidly, resulting in an estimated 343 thousand new US jobs between 2011 and 2015. Demand for US goods and services continues to climb throughout the period, adding an estimated $34 billion to US GDP in 2015, $40.4 billion in 2020, and $42.2 billion in 2025.[100]

These CERI estimates apply to the entire oil sands industry, however, not only the Keystone XLproject, and they are derived from a proprietary economic analysis which has not been subject to external review. Some stakeholders point to State Department and other studies reporting much lower anticipated economic benefits.[101] Consequently, it is difficult to determine what specific economic and employment impacts may ultimately be attributable to the Keystone XL pipeline. Nonetheless, given the physical scale of the project, it could be expected to increase employment and investment at least during construction.

Canadian Oil Sands Environmental Impacts

Oil production from oil sands is controversial because it has significant environmental impacts, including emissions of greenhouse gases during extraction and processing, disturbance of mined land, and impacts on wildlife and water quality.[102] Because bitumen in oil sands cannot be pumped from a conventional well, it must be mined, usually using strip mining or open pit techniques, or the oil can be extracted with underground heating methods.[103] Large amounts of water and natural gas are also required (for heating) during the extraction process.[104] The magnitude of the environmental impacts of oil sands production, in absolute terms and compared to conventional oil production, has been the subject of numerous, and sometimes conflicting, studies and policy papers.[105] Some stakeholders who object to oil sands projects oppose the Keystone XL pipeline because it expands access to new markets for the oil produced by those projects, thereby encouraging what they consider to be further environmentally destructive oil sands development. As discussed earlier, however, if oil sands production can be diverted to other markets (e.g., Asia), preventing the Keystone XL project may not necessarily limit oil sands development.[106]

Fossil Fuels Dependence and Greenhouse Gas Emissions

Some stakeholders object to the Keystone XL pipeline because it would increase U.S. supplies of oil, and thereby perpetuate the nation's dependence on imported fossil fuels and increase carbon emissions from the transportation sector.[107] Acknowledging this concern, in a public forum on October 20, 2010, Secretary of State Clinton reportedly remarked that "we're either going to be dependent on dirty oil from the [Persian] Gulf or dirty oil from Canada … until we can get our act together as a country and figure out that clean, renewable energy is in both our economic interests and the interests of our planet."[108] Critics of the State Department's draft and supplemental draft EIS assert that the environmental review overlooks the pipeline project's overall impact on greenhouse gas emissions, for example, from the extraction and refining processes. To address those potential emissions, EPA recommends that the final EIS include discussion of mitigation approaches for greenhouse gas emissions from extraction activities that are either currently used or could be employed to help lower lifecycle greenhouse gas emissions.[109] However, others have argued that whether the Keystone XL Pipeline is constructed would have little bearing on greenhouse gas emissions as there are likely to be other export routes available for Canadian oil sands crude, and therefore, the same crude oils would still be transported and refined, albeit in different locations.[110]

APPENDIX. PRESIDENTIAL PERMITTING AUTHORITY[111]

The executive branch has exercised permitting authority over the construction and operation of "pipelines, conveyor belts, and similar facilities for the exportation or importation of petroleum, petroleum products" and other products at least since the promulgation of Executive Order 11423 in 1968.[112] Executive Order 13337 amended this authority and the procedures associated with the review, but did not substantially alter the exercise of authority or the delegation to the Secretary of State in E.O. 11423.[113] However, the source of the executive branch's permitting authority is not entirely clear from the text of these Executive Orders. Generally, powers exercised by the executive branch are authorized by legislation or are inherent presidential powers based in the Constitution. E.O. 11423 makes no mention of any authority, and E.O. 13337 refers only to the "Constitution and the Laws of the United States of America, including Section 301 of title 3, United States Code."[114] Section 301 simply provides that the President is empowered to delegate authority to the head of any department or agency of the executive branch.

The legitimacy of this permitting authority has been addressed by federal courts. In *Sisseton v. United States Department of State*, the plaintiff Tribes filed suit and asked the court to suspend or revoke the Presidential Permit issued under E.O. 13337 for the TransCanada Keystone Pipeline.[115] The U.S. District Court for the District of South Dakota found that the plaintiffs lacked standing because they would be unable to prove their injury could be redressed by a favorable decision.[116] The court determined that even if the plaintiff's injury could be redressed, "the President would be free to disregard the court's judgment," as the case concerned the President's "inherent Constitutional authority to conduct foreign policy," as opposed to statutory authority granted to the President by Congress.[117]

The court further found that even if the Tribes had standing, the issuance of the Presidential Permit was a presidential action, not an agency action subject to judicial review under the Administrative Procedure Act (APA).[118] The court stated that the authority to regulate the crossborder pipeline lies with either Congress or the President.[119] The court found that "Congress has failed to create a federal regulatory scheme for the construction of oil pipelines, and has delegated this authority to the states. Therefore, the President has the sole authority to allow oil pipeline border crossings under his inherent constitutional authority to conduct foreign affairs."[120] The President could delegate his permitting authority to the U.S. Department of State, but delegation did not transform the permit's issuance into an agency action reviewable under the APA.[121]

In *Sierra Club v. Clinton*,[122] the plaintiff Sierra Club challenged the Secretary of State's decision to issue a Presidential Permit authorizing the Alberta Clipper pipeline. Among the plaintiff's claims was an allegation that issuance of the permit was unconstitutional because the President had no authority to issue the permits referenced in E.O. 13337 (in this case, for the importation of crude oil from Canada via pipeline).[123] The defendant responded that the authority to issue Presidential Permits for these border-crossing facilities "does not derive from a delegation of congressional authority ... but rather from the President's constitutional authority over foreign affairs and his authority as Commander in Chief."[124] The U.S. District Court for the District of Minnesota agreed, noting that the defendant's assertion regarding the source of the President's authority has been "well recognized" in a series of Attorney General opinions, as well as a 2009 judicial opinion.[125] The court also noted that these permits had been issued many times before and that "Congress has not attempted to exercise any exclusive authority over the permitting process. Congress's inaction suggests that Congress has accepted the authority of the President to issue cross-border permits."[126] Based on the historical recognition of the President's authority to issue these permits and Congress's implied approval through inaction, the court found the Presidential Permit requirement for border facilities constitutional.

End Notes

[1] This report provides an overview of the Keystone XL project, permit review process, and general policy issues. For more detailed legal analysis, see CRS Report R42124, *Proposed Keystone XL Pipeline: Legal Issues*, by Adam Vann et al. For more analysis of U.S.-Canada energy trade, see CRS Report R41875, *The U.S.-Canada Energy Relationship: Joined at the Well*, by Paul W. Parfomak and Michael Ratner.

[2] U.S. Department of State, "Keystone XL Pipeline Project Review Process: Decision to Seek Additional Information," November 10, 2011, http://www.state.gov/r/pa/prs/ps/2011/11/176964.htm.

[3] U.S. Department of State, "Briefing on the Keystone XL Pipeline," briefing transcript, January 18, 2012, http://www.state.gov/r/pa/prs/ps/2012/01/181492.htm.

[4] TransCanada Corp., "TransCanada Will Re-Apply for a Keystone XL Permit," press release, January 18, 2012.

[5] See, for example, Juliet Eilperin, "Democratic Lawmakers Pressure Obama Administration on Both Sides of Keystone Pipeline Issue," *Washington Post*, October 19, 2011; House Energy & Commerce Committee, Subcommittee on Energy and Power, Hearing on The American Energy Initiative, Discussion Draft of H.R. ____, the North American Made Energy Security Act of 2011, May 23, 2011; U.S. Senator Charles Grassley, Letter to Secretary of State Hillary Rodham Clinton, May 16, 2011; U.S. Senator Max Baucus, Letter to Secretary of State Hillary Rodham Clinton, September 10, 2010; U.S. Representative Henry A. Waxman, Letter to Secretary of State Hillary Rodham Clinton, July 2, 2010.

[6] U.S. Department of State, *Supplemental Draft Environmental Impact Statement for the Keystone XL Oil Pipeline Project*, April 15, 2011. p. 1-4. An initial capacity of 700,000 bpd may be raised to 830,000 bpd by increasing the pumping capacity. The Keystone XL project had applied to the Pipeline and Hazardous Materials Safety Administration to operate at slightly higher pressure than permitted in standard regulations, which would have enabled a 900,000 bpd capacity, but it withdrew its applications for such a Special Permit in August, 2010. The company may reapply for this exemption in the future, however, even after the pipeline is constructed, should it be approved.

[7] TransCanada Corp., "Media Advisory - State of Nebraska to Play Major Role in Defining New Keystone XL Route Away From the Sandhills," press release, November 14, 2011.

[8] TransCanada Keystone Pipeline, L.P., Application of TransCanada Keystone Pipeline L.P. for a Presidential Permit Authorizing the Construction, Operation, and Maintenance of Pipeline Facilities for the Importation of Crude Oil to be Located at the United States-Canada Border, U.S. Dept. of State, September 19, 2008, p. 10, http://www.keystonepipeline-xl.state.gov/clientsite/keystonexl.nsf/presidentialpermitapplication.pdf? OpenFileResource.

[9] "TransCanada Expects $1-Billion Cost Escalation for Keystone XL Pipeline," Canadian Press, February 17, 2011.

[10] Richard M. Pollastro et al., Assessment of Undiscovered Oil Resources in the Devonian-Mississippian Bakken Formation, Williston Basin Province, Montana and North Dakota, 2008, U.S. Geologic Survey, National Assessment of Oil and Gas Fact Sheet (2008–3021), April 2008, p. 1, http://pubs.usgs.gov/fs/2008/3021/pdf/FS08-3021_508.pdf. The Bakken formation also stretches into parts of Manitoba and Saskatchewan, Canada.

[11] Nathan Vanderklippe, "TransCanada to Move U.S. Crude Through Keystone," *The Globe and Mail*, January 26, 2011.

[12] For more on Bakken oil production, see CRS Report R42032, *The Bakken Formation: An Emerging Unconventional Oil Resource*, by Michael Ratner et al.

[13] Jeffrey Jones, "TransCanada plans U.S. Bakken pipeline link," *Reuters*, January 20, 2011.

[14] TransCanada, "TransCanada to Transport U.S. Crude Oil to Market Bakken Open Season a Success," press release, January 11, 2011, http://www.transcanada.com/5631.html.

[15] Vanderklippe, 2011.

[16] Enbridge, "Bakken Pipeline Project—Project Overview," press release, http://www.enbridge.com/BakkenPipelineProjects/BakkenPipelineProjectUS.aspx.

[17] Lauren Krugel, "TransCanada attracts support for Montana-to-Oklahoma crude pipeline," *The Canadian Press*, January 20, 2011.

[18] Selam Gebrekidan, "Bakken Rail Terminal Ships First Crude Cargo-Lario," *Reuters*, November 9, 2011.

[19] 15 USC §717f(c).

[20] See Executive Order 13337, "Issuance of Permits With Respect to Certain Energy-Related Facilities and Land Transportation Crossings on the International Boundaries of the United States," 69 *Federal Register* 25299, May 5, 2004, as amended, and Department of State

Delegation of Authority No. 118-2 of January 26, 2006. The source of Permitting Authority for relevant Executive Orders is discussed further in the **Appendix**.

[21] Executive Order 13337, at Sec. 1(g).

[22] In processing Presidential Permit applications, the State Department is also explicitly directed to review the project's compliance with the National Historic Preservation Act (16 U.S.C. §470f), the Endangered Species Act (16 U.S.C. §1531 et seq.), and Executive Order 12898 of February 11, 1994 (59 *Federal Register* 7629), concerning environmental justice. In processing the permit application for the Keystone XL Pipeline project, issues associated with NEPA compliance have drawn the most attention. In large part, that is likely because it is during the NEPA process that compliance with these, as well as any other environmental requirements, would be identified, documented, and demonstrated.

[23] U.S. Department of State press release, "Keystone XL Pipeline Project Review Process: Decision to Seek Additional Information," November 10, 2011, http://www.state.gov/r/pa/prs/ps/2011/11/176964.htm.

[24] 42 U.S.C. §4332(2)(C).

[25] U.S. Department of State, "Notice of Intent to Prepare an Environmental Impact Statement and to Conduct Scoping Meetings and Notice of Floodplain and Wetland Involvement and to Initiate Consultation under Section 106 of the National Historic Preservation Act for the Proposed TransCanada Keystone XL Pipeline," 74 *Federal Register* 5020, January 28, 2009.

[26] In preparing an EIS associated with a Presidential Permit, NEPA regulations promulgated by both the Council of Environmental Quality (CEQ) and the State Department would apply. CEQ regulations implementing NEPA (under 40 C.F.R. §§1500-1508) apply to all federal agencies. NEPA regulations applicable to State Department actions, which supplement the CEQ regulations, are found at 22 C.F.R. §161.

[27] For more analysis of NEPA requirements, see CRS Report RL33152, *The National Environmental Policy Act (NEPA): Background and Implementation*, by Linda Luther.

[28] Documents submitted by TransCanada are available online at http://www.keystonepipeline-xl.state.gov/clientsite/keystonexl.nsf?Open, under the heading "Project Documents."

[29] Any consultation or approval necessary to comply with any additional requirements should occur concurrently and be integrated with preparation of the EIS.

[30] 40 C.F.R. §1508.5. Also, Executive Order 13337 directs the Secretary of State to refer an application for a Presidential Permit to other specifically identified federal departments and agencies on whether granting the application would be in the national interest.

[31] For more information, see the U.S. Environmental Protection Agency's "Environmental Impact Statement (EIS) Rating System Criteria" at http://www.epa.gov/compliance/nepa/comments/ratings.html.

[32] Documents prepared by the U.S. Department of State related to its NEPA requirements are available online at http://www.keystonepipeline-xl.state.gov/clientsite/keystonexl.nsf?Open, under the heading "State Dept. Documents."

[33] U.S. Environmental Protection Agency's July 16, 2010 letter to the U.S. Department of State commenting on the draft EIS for the Keystone XL project is available at http://yosemite.epa.gov/oeca/webeis.nsf/%28PDFView%29/20100126/$file/20100126.PDF.

[34] See Secretary of State Hillary Clinton "Remarks on Innovation and American Leadership to the Commonwealth Club," San Francisco, CA, October 15, 2010, available at http://www.state.gov/secretary/rm/2010/10/149542.htm. The statement by Secretary Clinton was actually made in response to a question about the Alberta Clipper pipeline project which received a Presidential Permit from the State Department in 2009, a State Department spokesman later clarified that the Secretary was referring to the Keystone XL pipeline permit approval.

[35] For example, see the October 21, 2010 letter from Senator Mike Johanns to Secretary Clinton expressing his concern that her statement gave the appearance that approval of the pipeline

was a foregone conclusion http://johanns.senate.gov/public/?a=Files.Serve&File_id=8b090aa5-76fe-41ca-a674-ae9e37db8d36.

[36] U.S. Environmental Protection Agency's June 6, 2011 letter to the U.S. Department of State commenting on the supplemental draft EIS for the Keystone XL project is available at http://yosemite.epa.gov/oeca/webeis.nsf/%28PDFView%29/20110125/$file/20110125.PDF?OpenElement.

[37] Environmental analysis associated with pipeline project alternatives is provided in Volumes 1 and 2 of the final EIS, available on the U.S. Department of State's Keystone XL Pipeline Project website at http://www.keystonepipelinexl.state.gov/clientsite/keystonexl.nsf?Open.

[38] U.S. Department of State press release, "Keystone XL Final Environmental Impact Statement Released; Public Meetings Set," August 26, 2011, http://www.state.gov/r/pa/prs/ps/2011/08/171082.htm.

[39] These additional public meetings are not part of the NEPA process. Considering the strong public interest in the pipeline proposal (both opposed and in favor), the public hearings were part of the State Department's national interest determination.

[40] For example, see U.S. Department of State, *Record of Decision and National Interest Determination, TransCanada Keystone Pipeline, LP Application for Presidential Permit*, February 25, 2008, http://www.cardnoentrix.com/keystone/project/SignedROD.pdf.

[41] U.S. Department of State, November 10, 2011.

[42] U.S. Department of State, January 18, 2012.

[43] Chad Woodworth, "Nebraska Says Keystone XL Review to Take Months," *The Energy Daily*, January 26, 2012.

[44] U.S. House of Representatives, Energy and Commerce Committee, "Committee Approves Legislation to Increase North American Energy Production and Create Jobs," press release, June 23, 2011.

[45] See, for example, Representative Henry A. Waxman, "Opening Statement before the Full Committee Markup on Semi-Annual Committee Activity Report and H.R. 1938, the North American-Made Energy Security Act," June 23, 2011.

[46] U.S. Senator Bernard Sanders, et al., Letter to The Honorable Harold W. Geisel, Office of Inspector General, U.S. Department of State, October 26, 2011.

[47] See. for example, Elisabeth Rosenthal and Dan Frosch, "Pipeline Review is Faced With Question of Conflict," *New York Times*, October 7, 2011.

[48] Harold W. Geisel, United States Department of State, Office of Inspector General, "Information Memo for Deputy Secretary Burns," November 4, 2011. http://sanders.senate.gov/imo/media/doc/Special%20Review%20Keystone%20XL%20Pipeline%20Nov%202011 2.pdf.

[49] U.S. Department of State, "Keystone XL Pipeline Project Review Process: Decision to Seek Additional Information," Media Note, PRN 2011/1909, Office of the Spokesperson, November 10, 2011.

[50] The White House, Office of the Press Secretary, "Statement by the President on the State Department's Keystone XL Pipeline Announcement," November 10, 2011.

[51] U.S. Department of State, November 10, 2011.

[52] KETV NewsWatch 7, "Uncut: KETV's Rob McCartney Interviews President Obama," Omaha, NE, November 1, 2011, http://www.ketv.com/video/29652519/detail.html.

[53] U.S. Department of State, January 18, 2012.

[54] Ibid.

[55] Kerri-Ann Jones, Assistant Secretary of State for Oceans and International Environmental and Scientific Affairs, Testimony before the House Energy and Commerce Committee, Subcommittee on Energy and Power Hearing on the North American Energy Access Act, January 25, 2012.

[56] Jeff Wright, Director, Office of Energy Projects, Federal Energy Regulatory Commission, Testimony before the House Energy and Commerce Committee, Subcommittee on Energy and Power Hearing on the North American Energy Access Act, January 25, 2012.

[57] While the APA's definition of "agency" does not specifically exclude or include the president, the Supreme Court has held that exercises of presidential authority are not subject to judicial review because the president is not an agency (*Dalton v. Specter*, 511 U.S. 462, 470 (1994)). The Court has also held that the APA does not apply to the president based on separation of powers principles (*Franklin v. Massachusetts*, 505 U.S. 788, 800-01 (1992)).

[58] South Dakota Public Utilities Commission, Final Decision and Order; Notice of Entry Before the Public Utilities Commission of the State of South Dakota, In the Matter of the Application by Transcanada Keystone Pipeline, LP for a Permit Under the South Dakota Energy Conversion and Transmission Facilities Act to Construct the Keystone Pipeline Project, HP07-001, http://puc.sd.gov/commission/orders/HydrocarbonPipeline/2008/hp07-001.pdf.

[59] Montana Major Facility Siting Act, Title 75, Chapter 20.

[60] Nebraska Rev. Stat. §57-1101.

[61] Kevin O'Hanlon, "Governor Signs Two Bills Into Law," *Lincoln Journal Star*, November 23, 2011.

[62] TransCanada Keystone, L.P., *Keystone XL Project: Preliminary Environmental Report*, September 2008, Table 7, http://www.keystonepipeline-xl.state.gov/clientsite/keystonexl.nsf/preliminaryenvironmentalreport.pdf?OpenFileResource.

[63] TransCanada Corp., November 14, 2011.

[64] For more information, see the Department of the Interior's U.S. Fish and Wildlife Service web page on the Sand Hills at http://www.fws.gov/mountain-prairie/pfw/ne/ne4.htm.

[65] The entire Ogallala Aquifer system stretches across eight states generally from north to south to include South Dakota, Nebraska, Wyoming, Colorado, Kansas, Oklahoma, New Mexico, and Texas and underlies about 174,000 square miles.

[66] Generally, a release of crude oil to land would not necessarily result in groundwater contamination. In addition to the depth from the land surface to groundwater and the characteristics of the environment into which the crude oil is released (e.g., characteristics of the underlying soils), the potential for crude oil to reach groundwater would depend on factors such as the volume of the spill, the duration of the release, and the viscosity and density of the crude oil.

[67] TransCanada Keystone Pipeline, L.P., September 19, 2008, pp. 6-8.

[68] This is the case unless the oil is stranded due to transport bottlenecks. Ironically, the bottleneck for crude oil flowing south from the Midwest to the Gulf Coast—which Keystone XL would help alleviate—helped insulate Midwestern crude oil prices from the impacts of unrest in the Middle East and North Africa. However, as is discussed below, this may have benefited Midwestern refiners but probably did not significantly reduce costs for U.S. consumers.

[69] For more about this, see CRS Report R41683, *Middle East and North Africa Unrest: Implications for Oil and Natural Gas Markets*, by Michael Ratner and Neelesh Nerurkar.

[70] For a primer on the oil market, see CRS Video Brief *Introduction to the Oil Market*, at http://www.crs.gov/analysis/Pages/WVB00002.aspx.

[71] All data in this section are from the U.S. Energy Information Administration's (EIA's) *Petroleum Navigator* (http://www.eia.gov/dnav/pet/pet_move_impcus_a2_nus_ep00_im0_mbbl_m.htm) and *International Energy Statistics* (http://tonto.eia.doe.gov/cfapps/ipdbproject/IEDIndex3.cfm).

[72] For context, the United States consumes roughly 19 Mbpd, more than 20% of the world's oil market. Net imports are gross or total imports less total exports. This section will focus on gross imports, though it should be noted that among U.S. petroleum exports about 0.2 Mbpd of petroleum products go to Canada and 0.4 Mb/d to Mexico.

[73] These data are based on full year 2010 estimates provided by the EIA's *Short Term Energy Outlook* (STEO), http://www.eia.doe.gov/emeu/steo/pub/contents.html. The STEO provides a balance of U.S. supply and demand.

[74] For more analysis, see CRS Report R41765, *U.S. Oil Imports: Context and Considerations*, by Neelesh Nerurkar.

[75] U.S. Energy Information Administration, "Country Analysis Brief: Venezuela," February 2010, http://www.eia.doe.gov/emeu/cabs/Venezuela/Oil.html.

[76] For further analysis of U.S.-Canada energy trade, see CRS Report R41875, *The U.S.-Canada Energy Relationship: Joined at the Well*, by Paul W. Parfomak and Michael Ratner.

[77] As in the United States, Canadian consumption fell due to economic downturn. This allowed the increment in exports to be higher than the increment in production.

[78] Canadian Association of Petroleum Producers (CAPP), *Crude Oil: Forecast, Markets, and Pipelines*, June 2010, p. 2, http://www.capp.ca/getdoc.aspx?DocId=173003.

[79] CAPP, 2010, p. 7.

[80] CAPP, 2010, p. 13. According to CAPP, refineries adding capacity to process heavy oil in the Midwest include thosein Roxana, IL; Whiting, IN, and Detroit, MI.

[81] CAPP, 2010, p. 14.

[82] Baker Hughes, *Planning Ahead for Effective Canadian Crude Processing*, Baker Petrolite White Paper, 2010, http://www.bakerhughes.com/assets/media/whitepapers/4c2a3c8ffa7e1c3c7400001d/file/28271canadian_crudeoil_update_whitepaper_0610.pdf.pdf&fs=1497549.

[83] For a description of which units refineries may need to add (or have added) to be able to process more Canadian oil sands supply, see Praveen Gunaseelan and Christopher Buehler, "Changing US Crude Imports Are Driving Refinery Upgrades," *Oil and Gas Journal*, August 10, 2009.

[84] U.S. Department of State, April 15, 2011. p. 1-4.

[85] See increased U.S. crude oil production in the Midwest under the PADD2 heading at the following source: Energy Information Administration, U.S. Department of Energy, *Crude Oil Production (by PADD)*, Petroleum & Other Liquids, http://www.eia.gov/dnav/pet/pet_crd_crpdn_adc_mbblpd_a.htm.

[86] Adjusted for transport costs and other regional differences.

[87] Center for Energy Economics and Bureau of Economic Geology, *Overview of the Alberta Oil Sands*, University of Texas at Austin, 2006, p. 16, http://www.beg.utexas.edu/energyecon/documents/overview_of_alberta_oil_sands.pdf.

[88] For more about the U.S. refining system, see CRS Report R41478, *The U.S. Oil Refining Industry: Background in Changing Markets and Fuel Policies*, by Anthony Andrews, Robert Pirog, and Molly F. Sherlock.

[89] For an explanation of why the United States exports refined products when it is a large (crude) oil importer, see CRS Report R40120, *U.S. Oil Exports*, by Robert Bamberger.

[90] ConocoPhillips, "ConocoPhillips Pursuing Plan to Separate into Two Stand-Alone, Publicly Traded Companies," press release, July 14, 2011, http://www.conocophillips.com/EN/newsroom/news_releases/2011news/Pages/07-14- 2011.aspx.

[91] Enbridge, "Enbridge and Enterprise Agree to Reverse Seaway Crude Oil Pipeline From Cushing to U.S. Gulf Coast," press release, November 16, 2011, http://www.enbridge.com/MediaCentre/News.aspx?yearTab=en2011&id=1530773.

[92] Bradley Olson , "Enbridge Pursuing Alternative to Transcanada's Keystone XL," *Bloomberg*, November 9, 2011.

[93] Ben Lefebvre, "Enterprise Products Cancels Wrangler Pipeline," November 16, 2011.

[94] Lucretia Cardenas, "Kinder Morgan Says Eyes Fall Open Season For Trans Mountain Pipeline Expansion," *Platts*, March 24, 2011.

[95] Enbridge, "Northern Gateway at a Glance," press release, 2011, http://www.northerngateway.ca/project-info/northern-gateway-at-a-glance. The project would also include a pipeline to allow the import of 193,000 bpd of condensate, a light hydrocarbon that can be blended with bitumen to allow pipeline transport.

[96] Derrick Penner, "Opposition to Enbridge Northern Gateway pipeline grows," *Vancouver Sun*, December 2, 2010.

[97] Edward Welsch, "TransCanada: Oil Sands Exports Will Go to Asia if Blocked in U.S.," Dow Jones Newswires, June 30, 2010.

[98] EnSys Energy & Systems, Inc., *Keystone XL Assessment: Final Report*, Prepared for the U.S. Department of Energy, Office of Policy & International Affairs, December 23, 2010, p. 118.

[99] Energy Policy Research Foundation, Inc., The Value of the Canadian Oil Sands (….to the United States): An Assessment of the Keystone Proposal to Expand Oil Sands Shipments to Gulf Coast Refiners, Washington, DC, November 29, 2010, p. 2, http://www.eprinc.org/pdf/oilsandsvalue.pdf.

[100] Canadian Energy Research Institute, *The Impacts of Canadian Oil Sands Development on the United States' Economy, Final Report*, Calgary, Alberta, October 2009, p. vii.

[101] See, for example, Cornell University Global Labor Institute, *Pipe Dreams? Jobs Gained, Jobs Lost by the Construction of Keystone XL*, September 28, 2011; National Wildlife Federation, "TransCanada Exaggerating Jobs Claims for Keystone XL," November 9, 2010, http://www.dirtyoilsands.org/files/Keystone_XL_Jobs_11-09-10.pdf.

[102] For more analysis of oil sands and their environmental impacts, see CRS Report RL34258, *North American Oil Sands: History of Development, Prospects for the Future*, by Marc Humphries.

[103] U.S. Bureau of Land Management, "About Tar Sands," web page, January 11, 2011, http://ostseis.anl.gov/guide/tarsands/index.cfm.

[104] Cecilia Jamasmie, "The Challenges and Potential of Canada's Oil Sands," *Mining*, September-October 2010, pp. 7-8.

[105] For an example of contrasting views, see IHS CERA Inc., *Oil Sands, Greenhouse Gases, and US Oil Supply, Getting the Numbers Right*, 2010; and Natural Resources Defense Council, "Setting the Record Straight: Lifecycle Emissions of Tar Sands," November 2010.

[106] For more analysis of oil sands, including the environmental effects of its extraction, see CRS Report RL34258, *North American Oil Sands: History of Development, Prospects for the Future*, by Marc Humphries.

[107] See, for example: Natural Resources Defense Council, *Tar Sands Invasion: How Dirty and Expensive Oil from Canada Threatens America's New Energy Economy*, May 2010.

[108] See footnote 34.

[109] See EPA letter referenced in footnote 33, p. 7.

[110] EnSys Energy & Systems 2010, p. 116.

[111] For a more expansive treatment of this topic, see CRS Report R42124, *Proposed Keystone XL Pipeline: Legal Issues*, by Adam Vann et al.

[112] *Providing for the performance of certain functions heretofore performed by the President with respect to certain facilities constructed and maintained on the borders of the United States*, 33 Fed. Reg. 11741, August 16, 1968.

[113] *Issuance of Permits With Respect to Certain Energy-Related Facilities and Land Transportation Crossings on the International Boundaries of the United States*, 69 Fed. Reg. 25299, May 5, 2004.

[114] Ibid.

[115] 659 F. Supp. 2d 1071, 1078 (D. S.D. 2009).

[116] Ibid. at 1078.

[117] Ibid. at 1078, 1078 n.5.

[118] See ibid. at 1080-81.

[119] Ibid. at 1081.

[120] Ibid.

[121] Ibid. at 1082.

[122] 689 F.Supp.2d 1147 (D. Minn. 2010).

[123] Ibid. at 1162.

[124] Ibid.

[125] Ibid. at 1163 (citing 38 U.S. Atty Gen. 162 (1935); 30 U.S. Op. Atty. Gen. 217 (1913); 24 U.S. Op. Atty. Gen. 100; and Natural Resources Defense Council (NRDC) v. U.S.

Department of State, 658 F.Supp.2d 105, 109 (D.D.C. 2009)). The court in *NRDC* held that the State Department's issuance of a presidential permit under Executive Order 13337 was not subject to judicial review under the Administrative Procedure Act for abuse of discretion because "the issuance of presidential permits is ultimately a presidential action." 658 F. Supp. 2d at 109, 111-12. The court said that to allow judicial review of such decisions would raise separation of powers concerns. Ibid. at 111.

[126] Ibid.; see also Youngstown Sheet and Tube Co. v. Sawyer, 343 U.S. 579 (1952) (establishing a three-part test for analyzing the validity of presidential actions in relation to constitutional and congressional authority).

In: The Keystone XL Pipeline Project
Editors: G. B. Flynn and D. Burnett
ISBN: 978-1- 62081-221-1

Chapter 2

PROPOSED KEYSTONE XL PIPELINE: LEGAL ISSUES*

Adam Vann, Kristina Alexander, Vanessa K. Burrows and Kenneth R. Thomas

SUMMARY

In 2008, TransCanada Corp. applied for a presidential permit from the State Department to construct and operate an oil pipeline across the U.S.-Canada border in a project known as Keystone XL. The Keystone XL pipeline would transport oil produced from oil sands in Alberta, Canada, to Gulf Coast refineries. The permit application was subjected to review by the State Department pursuant to executive branch authority over cross-border pipeline facilities as articulated in Executive Order 13337.

After several phases of review, on November 10, 2011, the State Department announced that it would seek additional information about alternative pipeline routes before it could move forward with a national interest determination. In response, several pieces of legislation were introduced, including Title V of the Temporary Payroll Tax Cut Continuation Act of 2011. Title V dictated that President must grant the Keystone XL pipeline permit within 60 days of the law's enactment, unless the President

---

* This is an edited, reformatted and augmented version of the Congressional Research Service Publication, R42124, dated January 23, 2012.

determined that the pipeline is not in the national interest. If the President did not make a national interest determination and took no action to grant the permit, then the law provided that the permit "shall be in effect by operation of law." The Temporary Payroll Tax Cut Continuation Act of 2011 (P.L. 112-78), including Title V addressing the Keystone XL permit, was enacted on December 23, 2011.

Pursuant to the requirements of Title V, on January 18, 2012, the State Department recommended that "the presidential permit for the proposed Keystone XL pipeline be denied and, that at this time, the TransCanada Keystone XL Pipeline be determined not to serve the national interest." The same day, the President stated his determination that the Keystone XL pipeline project "would not serve the national interest."

New legislative activity with respect to the permitting of border-crossing facilities, a subject previously handled exclusively by the executive branch, has triggered inquiries as to whether this raises constitutional issues related to the jurisdiction of the two branches over such facilities. Additionally, as states have begun to contemplate taking action with respect to the pipeline siting, some have questioned whether state siting of a pipeline is preempted by federal law. Others argue that states dictating the route of the pipeline violates the dormant Commerce Clause of the Constitution which, among other things, prohibits one state from acting to protect its own interests to the detriment of other states.

This report reviews those legal issues. First, it suggests that legislation related to cross-border facility permitting is unlikely to raise significant constitutional questions, despite the fact that such permits have traditionally been handled by the executive branch alone pursuant to its constitutional "foreign affairs" authority. Next, it observes generally that state oversight of pipeline siting decisions does not appear to violate existing federal law or the Constitution. Finally, the report suggests that State Department's implementation of the existing authority to issue presidential permits appears to allow for judicial review of its National Environmental Policy Act determinations.

BACKGROUND

In September 2008, TransCanada Corp. (a Canadian company) applied to the U.S. Department of State for a permit to cross the U.S.-Canada international border with the Keystone XL pipeline project.[1] If constructed, the pipeline would carry crude oil produced from the oil sands region of Alberta, Canada, to U.S. Gulf Coast refineries. The U.S. portion of the Keystone XL pipeline project, as proposed, would pass through Montana, South Dakota, Nebraska, Oklahoma, and Texas. The pipeline would consist of approximately 1,380 miles of 36-inch-diameter pipe and have the capacity to transport 830,000 barrels per day of crude oil to the United States. The pipeline would deliver up to roughly 200,000 barrels per day to an existing oil terminal in Oklahoma, with the remainder sent further to points in Texas.[2]

In most instances, decisions about the siting of oil pipelines, even interstate oil pipelines like the proposed Keystone XL pipeline, are made by state governments if the state governments choose to exercise a pipeline siting authority.[3] The federal government generally does not regulate the siting of oil pipelines, although it does oversee oil pipeline safety[4] and pricing issues.[5] However, the construction, connection, operation, and maintenance of a pipeline that connects the United States with a foreign country requires the permission of the U.S. Department of State, conveyed through a presidential permit.[6] Accordingly, the proposed Keystone XL pipeline would require a permit. Executive Order 13337 delegates to the Secretary of State the President's authority to issue such a permit upon a determination that the project is in the national interest.[7]

Prior to making the national interest determination, the State Department conducts a review under the National Environmental Policy Act (NEPA).[8] For Keystone XL, a draft environmental impact statement (EIS) was issued April 16, 2010,[9] but the Environmental Protection Agency (EPA) found the draft was inadequate.[10] A supplemental draft was available in April 2011,[11] and again EPA found flaws.[12] A final EIS was announced in August of that year.[13] Under NEPA practice, the final EIS would be reviewed by EPA and other interested agencies prior to issuance of a record of decision.[14] However, a proposed route change through Nebraska put those final actions on hold.

On November 10, 2011, the State Department announced a decision to seek additional information about alternative pipeline routes before it could move forward with a national interest determination. Specifically, concerns regarding potential environmental impacts of constructing and operating the pipeline along the proposed route through the Sand Hills region of Nebraska led the State Department to decide that an assessment of potential alternative routes that would avoid that area was necessary.

On December 23, 2011, Congress passed and the President signed into law the Temporary Payroll Tax Cut Continuation Act of 2011. Title V of the act addressed the Keystone XL presidential permitting process.[15] Under that provision, the President was required to grant the Keystone XL pipeline permit within 60 days of the law's enactment, unless the President determined that the pipeline was not in the national interest.[16] If the President did not make a national interest determination and took no action to grant the permit, then the law provided that the permit "shall be in effect by operation of law."[17] Title V of the act appears to be the first legislative foray into the permitting of the border crossing facilities as contemplated in Executive Orders 11423 and 13337.

As required by Title V of the act, on January 18, 2012, the State Department recommended that "the presidential permit for the proposed Keystone XL pipeline be denied and, that at this time, the TransCanada Keystone XL Pipeline be determined not to serve the national interest."[18] The State Department asserted that its recommendation "was predicated on the fact that the Department does not have sufficient time to obtain the information necessary to assess whether the project, in its current state, is in the national interest."[19] The State Department press release also indicated that the 60-day time period provided for the in the act "is insufficient" for a determination as to whether the pipeline is in the national interest.[20] The State Department said that "subsequent permit applications" and "applications for similar projects" were not precluded by the denial of this particular permit application.[21] The same day, the President stated his determination that the Keystone XL pipeline project "would not serve the national interest."[22] He made this determination, as required by Title V of the act, in a memorandum to the Secretary of State.[23]

In addition to Title V of the Temporary Payroll Tax Cut Continuation Act of 2011, several other pieces of legislation have been introduced in Congress to compel expedited review of the Keystone XL application. The North American Energy Security Act (S. 1932), the American Energy Security Act (H.R. 3537), and the Middle Class Tax Relief and Job Creation Act of 2011 (H.R. 3630) also would require the Secretary of State to issue a permit for the project within 60 days of enactment, unless the President publicly determines the project to be not in the national interest. The January 18, 2012, determination that the project was not in the national interest may have superseded this proposed legislation. The North American Energy Access Act (H.R. 3548) would transfer permitting authority over the Keystone XL pipeline project from the State Department to the Federal Energy Regulatory Commission (FERC), and would require the commission to issue a permit for the project within 30 days of enactment. Also, other legislation may be introduced in the near future in response to the January 18, 2012, national interest determinations of the Secretary of State and the President.

The recent legislative activity appears to represent the first congressional efforts to amend the established executive branch procedure for the permitting of cross-border pipeline facilities, and there is some question as to whether this raises constitutional issues related to the jurisdiction of the two branches over such facilities. Additionally, as states consider taking action with respect to the pipeline siting, some have questioned whether state siting of a pipeline is preempted by federal law. Others have argued that states dictating the route of the pipeline violates the dormant Commerce Clause of the Constitution, which prohibits one state from acting to protect its own interests to the detriment of other states. Finally, some environmental groups have raised questions about the adequacy of government review of the proposed pipeline under the National Environmental Policy Act (NEPA), the routing of the pipeline near key aquifers, and the emissions caused by the oil production. This report reviews those legal issues, observing generally that legislation altering the pipeline border crossing approval process appears likely to be a legitimate exercise of Congress's constitutional authority to regulate foreign commerce and that state oversight of pipeline siting decisions does not appear to violate existing federal law or the Constitution. The report also suggests that the State Department's implementation of its authority to issue presidential permits appears to allow for judicial review of its NEPA determinations.

AUTHORITY TO ISSUE PERMITS FOR BORDER CROSSING FACILITIES: BALANCING EXECUTIVE AND LEGISLATIVE ROLES

Source of Presidential Authority to Regulate Foreign Commerce

The Constitution does not expressly accord the President any authority to regulate foreign commerce. However, the President's recognized authority in the area of foreign affairs permits him to take action in matters of foreign commerce such as border crossing facilities, as discussed below. There has been some judicial recognition of the President's ability to exercise authority over matters implicating foreign commerce, even in the absence of an express delegation of authority by Congress,[24] but the Supreme Court has not definitively opined on the circumstances in which any such authority may be exercised.

Source of Executive/State Department Permitting Authority

The executive branch exercises permitting authority over the construction and operation of "pipelines, conveyor belts, and similar facilities for the exportation or importation of petroleum, petroleum products" and other products pursuant to a series of executive orders. This authority has been vested in the U.S. State Department since the promulgation of Executive Order 11423 in 1968.[25] Executive Order 13337 amended this authority and the procedures associated with the review, but did not substantially alter the exercise of authority or the delegation to the Secretary of State in Executive Order 11423.[26]

Executive Order 11423 provided that, except with respect to cross-border permits for electric energy facilities, natural gas facilities, and submarine facilities,

> The Secretary of State is hereby designated and empowered to receive all applications for permits for the construction, connection, operation, or maintenance, at the borders of the United States, of: (i) pipelines, conveyor belts, and similar facilities for the exportation or importation of petroleum, petroleum products, coal, minerals, or other products to or from a foreign country; (ii) facilities for the exportation or importation of water or sewage to or from a foreign country; (iii) monorails, aerial cable cars, aerial tramways and similar facilities for the transportation of persons or things, or both, to or from a foreign country; and (iv) bridges, to the extent that congressional authorization is not required.[27]

Executive Order 13337 designates and empowers the Secretary of State to "receive all applications for Presidential permits, as referred to in Executive Order 11423, as amended, for the construction, connection, operation, or maintenance, at the borders of the United States, of facilities for the exportation or importation of petroleum, petroleum products, coal, or other fuels to or from a foreign country."[28] Executive Order 13337 further provides that after consideration of the application and comments received thereon: "[I]f the Secretary of State finds that issuance of a permit to the applicant would serve the national interest, the Secretary shall prepare a permit, in such form and with such terms and conditions as the national interest may in the Secretary's judgment require, and shall notify the officials required to be consulted ... that a permit be issued."[29] Thus, the Secretary of State is directed by the President to authorize all border crossing facilities that the Secretary has determined would "serve the national interest."

However, the source of the executive branch's permitting authority is not explicitly stated within the executive orders. Powers exercised by the executive branch are authorized by legislation or are inherent presidential powers based in the Constitution. Executive Order 11423 does not reference any statute or constitutional provision as the source of its authority, although it does state that "the proper conduct of foreign relations of the United States requires that executive permission be obtained for the construction and maintenance" of border crossing facilities.[30] Executive Order 13337 refers only to the "Constitution and the Laws of the United States of America, including Section 301 of title 3, United States Code."[31] However, 3 U.S.C. Section 301 simply provides that the President is empowered to delegate authority to the head of any department or agency of the executive branch.

Federal courts have addressed the legitimacy of this permitting authority. In *Sisseton-Wahpeton Oyate v. U.S. Department of State*, the plaintiff Tribes asked the court to suspend or revoke a presidential permit issued under Executive Order 13337 for the TransCanada Keystone Pipeline.[32] The plaintiffs claimed that the issuance of the national interest determination and the border crossing permit for the project violated NEPA and the Administrative Procedure Act (APA). The U.S. District Court for the District of South Dakota determined that even if the plaintiffs' injury could be redressed, "the President would be free to disregard the court's judgment," as the case concerned the President's "inherent constitutional authority to conduct foreign policy," as opposed to statutory authority granted to the President by Congress.[33] The court further found that even if the Tribes had standing, the issuance of the presidential permit was a presidential action, not an agency action subject to judicial review under APA.[34]

The court stated that the authority to regulate the cross-border pipeline lies with either Congress or the President.[35] The court found that "Congress has failed to create a federal regulatory scheme for the construction of oil pipelines, and has delegated this authority to the states. Therefore, the President has the sole authority to allow oil pipeline border crossings under his inherent constitutional authority to conduct foreign affairs."[36]

In *Sierra Club v. Clinton*,[37] the plaintiff Sierra Club challenged the Secretary of State's 2009 decision to issue a permit authorizing a pipeline from Alberta, Canada, to Superior, Wisconsin. The plaintiff claimed that issuance of the permit was unconstitutional because the President had no authority to issue the permits referenced in Executive Order 13337.[38] The defendant responded that the authority to issue permits for these border-crossing facilities "does not derive from a delegation of congressional authority ... but rather from the President's constitutional authority over foreign affairs and his authority as Commander in Chief."[39] The U.S. District Court for the District of Minnesota agreed, noting that the defendant's assertion regarding the source of the President's authority has been "well recognized" in a series of Attorney General opinions, as well as a 2009 judicial opinion.[40] The court also noted that these permits had been issued many times before and that "Congress has not attempted to exercise any exclusive authority over the permitting process. Congress's inaction suggests that Congress has accepted the authority of the President to issue cross-border permits."[41] Based on the historical recognition of the President's authority to issue those permits and Congress's implied approval through inaction, the court found the permit requirement for border facilities constitutional.

As these cases show, courts that have analyzed the President's exercise of permitting authority as articulated in Executive Order 13337 have held that it is a legitimate exercise of the President's constitutional authority, and therefore does not require legislative authorization.

Source of Congressional Authority to Regulate Foreign Commerce

Article I, Section 8 of the Constitution authorizes Congress to "regulate Commerce with foreign Nations." Whereas any independent presidential authority in matters affecting foreign commerce derives from the President's more general foreign affairs authority, Congress's power over foreign commerce is plainly enumerated by the Constitution, suggesting that its authority in this field is preeminent. In a review of the origins of the Constitution's Foreign Commerce Clause, the former Special Counsel of the Department of Justice's Office of Legal Counsel emphasized the placement of the foreign commerce power with Congress, stating that

> the power to regulate foreign commerce at the national level was to be vested in Congress. ... The debate at the Philadelphia Convention over whether a bare majority or a supermajority of each House was required to enact foreign commerce regulations demonstrates that the Framers intended such regulation to be made by a legislative body, rather than an executive or judicial one.[42]

Reconciling the Executive and Legislative Roles Related to Foreign Commerce and Border Crossing Facilities

Each of the three branches of the federal government has opined on the nature of the executive and legislative powers related to matters related to foreign affairs and foreign commerce, in many cases addressing border crossing facilities specifically.

Executive Branch Deference to Congress on Cross-Border Facilities

Until recently, Presidents have indicated that executive permission for cables, pipelines, and other border crossing facilities may be granted or refused only to the extent that Congress has not acted to institute terms and conditions on such facilities. For example, in 1869, President Grant resisted the landing of a transatlantic cable from France for which the French government had granted a monopoly.[43] President Grant recounted his unwillingness to act "[i]n the absence of legislation by Congress," but stated that he dropped his resistance to the cable once the monopolistic condition was removed.[44] Letters indicate that he understood that Congress could impose "limitations and conditions" upon the laying of cables within U.S. waters.[45] He then proposed legislation to Congress with conditions "for the protection of the public" from monopolies with the right to operate cable telegrams.[46] He also indicated that "unless Congress otherwise direct[ed]," he would not oppose the landing of any other cable that complied with the conditions he had outlined and presented to Congress.

As another example, in 1897, President McKinley declined to act on request for an application from a French company to the State Department, requesting permission to land a supplementary cable under the "same terms and conditions as those which were imposed by the President in 1879 when the original cable was landed."[47] The State Department's letter to the French Ambassador stated that the President did "not regard himself as clothed, in the absence of legislative enactment, with the requisite authority to take any action upon the application."[48] President McKinley's position stemmed from the congressional introduction of a bill to grant the President "express authority to authorize the landing of submarine cable … subject to conditions therein specified," which "failed to become law."[49]

A series of Attorney General opinions dating back to the 19th century also suggest that the executive branch has taken an expansive view of Congress's authority in the area of cross-border commercial activity. These Attorney General opinions conditioned the President's authority to prohibit or permit cross-border cables, telegraphs, and electrical power on the absence of congressional action. The Department of Justice has cited these Attorney General opinions as "indicat[ing] that the President or his delegates for over a century have exercised the inherent executive authority to issue such permits without action by Congress."[50] These opinions assert that the President may permit or prohibit the landing of a foreign cable, subject to legislation by Congress; recognize that cross-border cables are "instrument[s] of commerce"; and indicate that the President's power to act in the absence of legislation on point derives from his "power to prohibit [the landing of a foreign cable] should he deem it an encroachment on our rights or prejudicial to our interests."[51]

The first of these Attorney General opinions, dating to 1898, formed the basis for future Attorney General opinions on submarine cables, wireless telegraphy, and gas pipelines. The Solicitor General, acting as the Attorney General, addressed the Secretary of State's questions as to "the power of the President, in the absence of legislative enactment, to control the landing of foreign telegraphic cables."[52] The acting Attorney General reviewed presidential actions and related communications regarding the landing of submarine cables dating back to the landing of the first foreign cable from Cuba to Florida in 1867.[53] The first cable to the United States landed "under the supposed authority" of a 1866 congressional act granting a monopoly to one company, but reserving to Congress "the power to alter and determine the rates."[54] However, Congress had not always enacted legislation imposing terms and conditions on subsequent cables that came to the shores of the United States. Presidents in the latter half of the 19th century granted permission for the landing of several cables, while indicating that the executive branch's permission in such matters was "subject to future action by Congress."[55]

In this 1898 opinion, the Attorney General stated that the President was charged with the "preservation of our territorial integrity and the protection of our foreign interests."[56] The Attorney General found that the President's ability to act was not limited to enforcing particular congressional acts, but that he was under oath to "preserve, protect, and defend the Constitution."[57] The opinion relied on the President's position as "head of the diplomatic service" and Commander-in-Chief to find that the President could impose conditions upon foreign cables, in absence of congressional legislation, and either prevent or permit the landing of such cables "on conditions which will protect the interests of this Government and its citizens."[58]

Subsequent Attorney General opinions reached a similar conclusion. For example, 19th century legislation directed that the United States not direct any "franchises, or concessions of any kind whatever" in Cuba while the United States occupied the island.[59] In an 1899 opinion, the Attorney General indicated that the executive branch had regulated the matter of cables, "revocable either at the will of the President or by subsequent legislation by Congress."[60] The general in charge of U.S. forces in Cuba had forbidden "grants and concessions of franchises" in Cuba, unless the Secretary of War approved.[61] Although the Attorney General did not "concede that Congress, by legislative act, has the power to restrain or control the proper exercise of the powers of the Commander in Chief of the Army and Navy of the United States, occupying, … a foreign territory," the Attorney General found that the expressed congressional will conformed to the executive branch's previous policies and was "entitled to the respect of the Executive Departments, and ought to be followed."[62] The Attorney General thus advised the Secretary of War "that it would be inexpedient … to grant permission to the applicant in this case to land its cable upon the soil of Cuba."[63]

Similarly, a 1913 Attorney General opinion interpreting a 1910 treaty between the United States and Great Britain concerning the diversion of water from the Niagara River for power purposes, as well as a statute on a similar subject, suggests that both the President and Congress have authority to regulate such international commercial transactions.[64] The opinion found that it was the President's "duty to interfere with any diversion within the State of New York of the waters of the Niagara River … for power purposes, exceeding a daily maximum" established in the treaty.[65] The Attorney General said that the treaty's stipulation that "the United States may authorize and permit" the diversion of waters up to a daily maximum "clearly places the subject, in the absence of legislation by Congress to the contrary, under [the President's] supervision."[66]

The opinion also responded to the President's question as to whether he had "any authority to control the importation into [the United States] from Canada of electric current generated by water power from the Niagara River."[67] The Attorney General opined that, due to the lapse of a statute regarding electricity transmission from Canada, the President was "free to control the matter under [his] plenary power to prevent any physical connection (not authorized by Congress) between any foreign country and the United States."[68] The opinion recognized the executive branch's power as being "subject, of course, to affirmative control by Congress" to prohibit or permit the importation of electrical power into the United States from Canada.[69] The Attorney General also concluded that "in the absence of legislation by Congress," the President could subject the importation of Canadian electrical power "to such conditions as to [him] may seem good."[70]

Finally, it should be noted that Executive Order 11423, discussed earlier in this report, delegated authority to the Secretary of State to receive all applications for permits for cross border facilities including pipelines, cable cars, bridges, and similar facilities only "to the extent that congressional authorization is not required."[71] However, Executive Order 13337 omitted the qualification "to the extent that congressional authorization is not required," when amending the section in Executive Order 11423 that empowered the Secretary to receive all such applications.[72]

Congressional Action Related to Foreign Commerce

A history of legislation related to border crossing facilities further suggests that congressional action related to permitting of pipeline border crossings is a legitimate exercise of Congress's authority to regulate foreign commerce. Examples of congressional legislation "regarding certain types of border crossing facilities," include the Submarine Cable Landing Licensing Act of 1921 and the International Bridge Act of 1972.[73] Additionally, Congress has passed several statutes relating specifically to petroleum or pipelines, and some of these require the President to make particular findings, such as that certain exports are in the national interest: the Export Administration Act of 1979,[74] the Mineral Leasing Act of 1920,[75] the Naval Petroleum Reserve Production Act of 1976,[76] the Outer Continental Shelf Lands Act Amendments of 1978,[77] the Energy Policy and Conservation Act,[78] and the Comprehensive Anti-Apartheid Act of 1986.[79]

Judicial Interpretations of the Executive and Legislative Authorities to Regulate Foreign Commerce

The Supreme Court's pronouncements on the foreign commerce clause recognize the expansive scope of this enumerated congressional power.[80] The Court has said that this power is "exclusive and plenary."[81] In the seminal 1824 Commerce Clause case *Gibbons v. Ogden*, the Supreme Court said the words of the Commerce Clause "comprehend every species of commercial intercourse between the United States and foreign nations. No sort of trade can be carried on between this country and any other, to which this power does not extend."[82] As of 2006, the Supreme Court had "never struck down an act of Congress as exceeding its powers to regulate foreign commerce."[83]

In addition, although there is no precedent that directly addresses Congress's power to legislate pipeline border crossings, language found in multiple judicial opinions further suggests that the executive authority to regulate such facilities may be amended by legislation. As noted above in *Sierra Club v. Clinton,*[84] the plaintiff challenged the Secretary of State's 2009 decision to issue a permit authorizing a pipeline from Canada to Wisconsin. The reviewing federal district court found that the authority to issue permits for these border-crossing facilities "does not derive from a delegation of congressional authority ... but rather from the President's constitutional authority over foreign affairs and his authority as Commander in Chief."[85] In reaching its conclusion, the court noted that "despite the fact that cross-border permits for pipelines have been issued by Presidents in the past, Congress has not attempted to exercise any exclusive authority over the permitting process. Congress's inaction suggests that Congress has accepted the authority of the President to issue cross-border permits."[86] This seems to suggest a recognition by the court that the cross-border permitting process is not the exclusive province of the executive branch and that Congress may legislate in this area if it is so inclined.

Similarly, in *Sisseton-Wahpeton Oyate v. U.S. Department of State,*[87] the reviewing federal district court engaged in a discussion of the nature of the authority being exercised by the President in granting cross-border permits. The court noted that

> In this case, the proposed pipeline crosses international borders. *Under the federal Constitution, then, the authority to regulate such a project vests in either the legislative or executive branch of government.* Congress has failed to create a federal regulatory scheme for the construction of oil pipelines, and has delegated this authority to the states. Therefore, the President has the sole authority to allow oil pipeline border crossings under his inherent constitutional authority to conduct foreign affairs.[88]

As with the Attorney General opinions discussed above, the court's language suggests that Congress possesses the constitutional authority to legislate border crossing facilities, which suggests that Congress also can legislate to amend the role of the executive branch with respect to such facilities.

Summary

As the discussion above demonstrates, all three branches of the federal government have historically taken an expansive view of the nature of Congress's authority to regulate foreign commerce, even in the absence of existing legislation in the area. Although the Temporary Payroll Tax Cut Continuation Act of 2011 appears to be the first legislation enacted on the subject of cross border pipeline permitting, the absence of previous legislation related to the permitting of cross border facilities does not mean that Congress lacks the constitutional authority to take action on these matters.

The executive branch also possesses some ability to act in the area of border crossing permitting, derived from the power to conduct foreign affairs under Article II of the Constitution. The executive's ability to act in this area, however, is informed by the previous lack of federal legislation in this area. The absence of legislation up to this point may have "enable[d], if not invite[d], measures on independent presidential responsibility" in which the President has acted in the "absence of either a congressional grant or denial of authority."[89] However, if Congress chose to assert its authority in the area of border crossing facilities, this would likely be considered within its constitutionally enumerated authority to regulate foreign commerce.[90] Congress may consider legislation to overturn the domestic effect of legal action denying a permit for a border crossing facility for the Keystone XL pipeline. It could also potentially establish criteria for the issuance of any cross border permits, and potentially require the issuance of permits to entities which fulfill such criteria.

CONSTITUTIONAL CONCERNS RELATED TO POTENTIAL ACTION BY STATES

As noted earlier, the federal government's role in siting the Keystone XL pipeline limited to the border crossing facility, although its environmental review considered impacts of the entire project pursuant to NEPA. However, some state and local government officials voiced concern regarding the pipeline's proposed route through areas perceived to present safety or environmental issues. At least one state, Nebraska, took action in November 2011, adopting two laws that (1) carve out a role for the state in siting "major oil pipelines" that run through the state,[91] and (2) directing collaboration with the State Department on a supplemental environmental statement for the proposed pipeline.[92]

Debate over the legislation in Nebraska gave rise to two constitutional concerns that some feel may limit the extent to which states may legislate and regulate interstate oil pipelines generally, and the proposed Keystone XL pipeline in particular. First, some have expressed concern that a state action may place too great of a burden on interstate commerce, in violation of a doctrine known as the dormant Commerce Clause. Second, some have wondered whether state legislation or regulation of interstate pipelines is preempted by federal legislation related to interstate pipeline safety.

The Dormant Commerce Clause

Legal Background

The Constitution provides that Congress shall have the power to regulate commerce with foreign nations and among the various states.[93] This power has been cited as the constitutional basis for a significant portion of the laws passed by Congress over the last 50 years, and it currently represents one of the broadest bases for the exercise of congressional authority. Although the Constitution does not explicitly provide that this federal power displaces the power of states to regulate interstate commerce, the courts have long found that such restrictions are an inherent part of the federal authority.[94]

State interference with trade was widespread before the ratification of the Constitution and a source of much dispute, so one goal of granting Congress the power to regulate interstate commerce was to limit such practices.[95] It was not clear at first to what extent states retained authority over such commerce. However, the doctrine of a dormant Commerce Clause was soon recognized by the courts as a limit on state legislation, even in the absence of congressional regulation.[96] *Dormant* is used to describe this right because it is not an express provision in the Constitution.

Although the meaning and application of the dormant Commerce Clause has varied over time, the modern standard, as applied to state regulatory actions, contains two basic inquiries. First is the question of whether a state regulation discriminates against other states on its face or in effect, showing an intent to benefit in-state economic interests at the expense of out-of-state interests. If that is the case, then no further analysis is required,[97] and the law will generally be struck down.[98] Second, if a law is non-discriminatory, but still has some impact on interstate commerce, the court will evaluate the law using a balancing test to find whether there is a legitimate public purpose.[99] State public health, environment and safety concerns are generally considered by the courts to represent a legitimate public purpose.

State Authority for Energy Facility Siting

As a preliminary matter, it should be noted that land use regulation, such as zoning, has traditionally been considered primarily a state concern. Also, there are many instances of states taking an important role in regulating the siting of energy facilities and associated rights-of-way. This includes state statutes governing the siting of oil pipelines within their borders.[100] In addition, siting electricity transmission lines has also historically been the exclusive province of states regardless of the nature of the facilities and the degree to which the interstate electricity grid depends upon those facilities. States retain primary authority to make siting decisions even in areas considered particularly important to the interstate electricity grid.[101] Other federal legislation has recognized the state's role in making siting decisions with respect to other tools of interstate commerce like wireless telecommunications facilities[102] and interstate highways.[103]

Of course, federal legislation may reduce the primacy of state authority. For instance, federal law provides that siting decisions for interstate natural gas pipelines are made by FERC pursuant to Section 7 of the Natural Gas Act.[104] Further, interstate natural gas pipelines still must obey state and local laws, including zoning ordinances, unless preempted by federal law. Thus, there seems to be little question that both historical practice and modern regulatory schemes recognize a significant role for state regulatory actions regarding siting energy facilities and associated rights-of-way.

Application of the Dormant Commerce Clause to State Action Related to Oil Pipeline Siting

In order to make a dormant Commerce Clause analysis, the court must have a fully developed record to allow "a sensitive, case-by-case analysis of purposes and effects."[105] Such a record is lacking in the Keystone XL case thus far. Without knowing what final action states may take, it is difficult to speculate which, if any, of those potential actions might present dormant Commerce Clause concerns. It is possible, however, to outline the factors a court might consider in reviewing a dormant Commerce Clause challenge to state action related to the pipeline.

The first question that a court would consider in a dormant Commerce Clause challenge would be whether a particular state regulation or siting decision was discriminatory. So long as the regulatory scheme provided a uniform standard for reviewing siting decisions for energy transmission it seems unlikely it would be found unconstitutional due to discrimination. In order to be found discriminatory, it would need to be shown that the intent or effect of the legislation was to benefit in-state economic interests at the expense of out-of-state interests. For instance, protectionist legislation which favored in-state companies regarding right-of-way acquisition might run afoul of this requirement. However, to the extent that the law applies equally to both instate and out-of-state pipeline companies, it is not clear that there would be a basis for an argument that such a law was discriminatory.[106]

If a law, statute or siting decision was found not to be discriminatory, then a court would move to a *Pike* analysis, in which an excessive burden on interstate commerce must be shown. One court described the *Pike* analysis as putting the burden on the challenging party to show that a statute encumbers interstate commerce in a way that "is clearly excessive in relation to the putative local benefits."[107] Although lower courts have varied in their application of this test,[108] it does appear that states are given significant deference by courts to establish environmental, public health, and safety standards.

Some have posited that a court might apply a heightened level of scrutiny to any state action regarding the Keystone XL pipeline on the grounds that the action would regulate "foreign commerce." The Supreme Court has held that state action which burdens foreign commerce should be subjected to a "more rigorous and searching scrutiny."[109] However, this heightened level of scrutiny has generally been applied to those areas where the state action directly affects foreign trade or foreign entities, not where it burdens commerce generally. Accordingly, it is unlikely that any state action related to the siting of the Keystone XL pipeline within its borders would be seen as a direct regulation of foreign trade. Rather, the action would most likely be seen as dictating the terms and conditions of the domestic transit of an imported good after it has been imported.

Next, putative local benefits would be identified and weighed. It does not appear to be difficult to establish before a court that an oil pipeline siting statute would serve local interests. As discussed above, courts have considered safety and economic concerns in evaluating similar legislation, and a court might also consider environmental, aesthetic, or any number of other non-protectionist concerns.[110] The relative weight of the local benefits would depend not only on the benefits identified by a court, but also on what implementing regulations look like and how a regulatory scheme is applied in a particular case.

Once a court had established what legitimate state interest was being served, it would then evaluate the level of burden on interstate commerce that was imposed by the state regulatory activity. In general, it appears unlikely that the passage of an oil siting statute similar to laws in other states would by itself create an "undue burden on interstate commerce" that would dictate reversal of the law. Statutes that require state permission prior to construction and operation and conformity with siting and regulatory requirements as set forth by a state agency during the review process are not uncommon and appear to be challenged only rarely.[111]

A state action related to the siting of an oil pipeline could be found to impose a significant burden on commerce. For example, if the statute requires the owner/operator of a proposed pipeline to obtain a permit before siting a pipeline in the state, but the regulatory agency tasked with enforcing the statute refused to permit a proposed pipeline across the state regardless of location, that refusal arguably may constitute an undue burden on interstate commerce.[112] However, even that decision might not be a *per se* violation of the dormant Commerce Clause, as a court might choose to evaluate the burden imposed by the re-routing of the pipeline through other states, or the burden imposed by using an alternative means of transport of the oil.

Assuming that a state statute or regulatory scheme would not preclude the passage of the Keystone XL pipeline through the state, it may still impose some administrative burden, and, in addition, might delay project construction or make the project more expensive. The burden on commerce is measured by the level of burden, delay, and expense. A court would consider what costs could be associated with this delay, and how this would affect the overall viability of the project. As noted, however, the fact that there are costs associated with regulatory burdens does not appear to have led to significant challenges to oil pipeline siting statutes in the past.

One argument that could be made is that new state requirements for oil pipelines in the Keystone XL pipeline planning process where none previously existed could add significantly to the cost or viability of the pipeline. Establishing the economic effect of a delay would depend on factors that could not be easily established until legislation was passed and siting decisions were made. For instance, what changes to the Keystone XL planning process would be required because of a change in siting to the original proposal? What costs associated with the pipeline have already been expended, and would the value of those expenditures be diminished by failure to begin construction by a certain date? What financial commitments might expire, and what would be the costs of new ones? How might delay otherwise raise costs, and would the economic viability of the project be threatened? What other economic or business considerations would be involved?

One could argue further that a court would need to consider how changed regulations might affect the burden on commerce. Thus, to the extent that the Keystone XL Pipeline planning process had accounted for certain regulatory hurdles, the state action might not significantly add to those regulatory burdens. However, new regulations could mean additional processing. To the extent that a court found that delays, even if costly, were a necessary component of the state effectuating legitimate state interests in public health and safety, then a court would be likely to find that the state's action was not an unconstitutional burden on interstate commerce.

Federal Preemption of State Pipeline Legislation

Background

Before turning to whether and how a state statute might be preempted, it is useful to review the history and content of federal pipeline safety legislation. The initial federal legislation governing fossil fuel pipeline safety was the Natural Gas Pipeline Safety Act of 1968.[113] This pipeline safety legislation was amended in 1979 to include liquid fuels,[114] and has been since been amended on a number of occasions, including by the Pipeline Safety Improvement Act of 2002.[115] Thus, what is often referred to as the "Pipeline Safety Act" is actually several statutes enacted over the last fifty years and codified at 49 U.S.C. Section 60101 *et seq*. However, for the sake of convenience this memorandum will refer to the collection of federal legislation codified at 49 U.S.C. Chapters 601-605 as the "Pipeline Safety Act."

How Preemption Is Evaluated

In addition to the arguments that state action regulating interstate pipelines might run afoul of the dormant Commerce Clause, several parties have also voiced concerns as to whether certain state actions related to oil pipelines could be preempted by the federal Pipeline Safety Act. Under the Supremacy Clause,[116] statutes and treaties as well as the constitution itself supersede state laws that "interfere with, or are contrary to" their dictates.[117] In such cases, the federal law is said to "preempt" the state law. In evaluating Supremacy Clause challenges to state statutes or other state actions, courts will generally consider three ways that a state action might be preempted. First, when acting within its constitutional limits, Congress may expressly preempt state law.[118] Second, federal law may preempt state law if the two conflict.[119] This type of conflict occurs either when "compliance with both federal and state regulations is a physical impossibility"[120] or if the state law "stands as an obstacle to the accomplishment and execution of the full purposes and objectives of Congress."[121] And third, "Congressional intent to preempt state law in a particular area may be inferred where the scheme of federal regulation is sufficiently comprehensive to make reasonable the inference that Congress 'left no room' for supplementary state regulation."[122]

The preemption argument in the case of Keystone XL is based on a concern that potential state action affecting pipeline siting might be preempted by federal pipeline safety statutes, despite state intent to protect public safety and the environment.

Express Preemption of State Action Related to the Proposed Keystone XL Pipeline

Turning first to express preemption, the Pipeline Safety Act contains the following language:

> A State authority that has submitted a current certification under section 60105(a) of this title may adopt additional or more stringent safety standards for intrastate pipeline facilities and intrastate pipeline transportation only if those standards are compatible with the minimum standards proscribed under this chapter. *A state authority may not adopt or continue in force safety standards for interstate pipeline facilities or interstate pipeline transportation.*[123]

In order to determine what state action is expressly preempted, it must first be determined what is meant by "safety standards for interstate pipeline facilities or interstate pipeline transportation." It is difficult to draw specific parameters as to what sort of state action would constitute a "safety standard for interstate pipelines facilities or interstate pipeline transportation" and therefore would be an area in which any state action would be expressly preempted by the Pipeline Safety Act. There are at least five sources for guidance on the parameters of the Pipeline Safety Act's express preemption: related statutory provisions, the "plain meaning" of the statutory language, the "industry usage" of the language, the federal agency's exercise of the statutory authority granted to it, and interpretations of the language by the federal courts.

Related Statutory Provisions

The Pipeline Safety Act offers an express statement of federal preemption, but also includes a limitation on the federal exercise of authority pursuant to the act: "[t]his chapter does not authorize the Secretary of Transportation to prescribe the location or routing of a pipeline facility."[124] This suggests that so long as the state action can be deemed only to "prescribe the location or routing of a pipeline facility," there would likely be no federal preemption.

The difficulty for proposed state actions affecting interstate pipelines arises where decisions related to location and routing may impact matters related to safety. This analysis should provide guidance to the state in determining what action can be taken without crossing the line from a legitimate state action affecting location/routing to a state adoption of "safety standards" that would be preempted by the Pipeline Safety Act.

Before moving forward with this analysis of what action may avoid being deemed preempted, it is important to note that the Pipeline Safety Act's preemption provision prohibits states from adopting safety standards for both "interstate pipeline facilities" and for "interstate pipeline transportation."[125] It is not clear what sort of state action might be preempted by a prohibition on safety standards for "pipeline facilities" and "pipeline transportation," but it is important that a state be aware of the existence of this prohibition and note that a court would likely seek to address this language in any preemption challenge based on 49 U.S.C. Section 60104(c).

The Plain Meaning of the Preemption Clause Related to Safety Standards

The primary rule of statutory interpretation is that, in the absence of a statutory definition (as is the case here), words in statutes are to be given their "plain meaning," and the plain meaning of "safety standards" is generally understood to be explicit rules regarding levels of safety that must be attained.

It is unclear whether the plain meaning or ordinary usage of the phrase "safety standards" would extend beyond pipeline performance standards to action taken by a state with respect to the siting of pipelines- an authority that is expressly *not* preempted by the Pipeline Safety Act- that is done for the express purpose of protecting public safety.[126] If however the state action related to siting were made for reasons not related to safety, a "plain meaning" reading of the statute would still suggest that there would be no preemption.

Industry Usage of the Phrase "Safety Standards"

A second rule of statutory interpretation is that some words or phrases may be ascribed the meaning given to them in the field addressed by the statute. The industry's perception of what concepts are (and are not) considered aspects of pipeline "safety" is therefore instructive here. One such example are the "safety and government-cited standards" for "pipeline operations" published by the American Petroleum Institute (API).[127] Of those standards and recommendations, only "managing system integrity" seems to relate to the pipeline's location, and, on initial review, those standards seem to be concerned only with different standards and practices that may apply in certain "high consequence" areas, rather than planning or routing concerns.

The API also has "safety and fire prevention" standards which seem to contemplate emergency prevention and response rather than external safety issues.[128] These industry-sourced standards may suggest that the industry considers "safety standards" to include only standards that impact pipeline operations such as avoiding leaks, preparedness and early warning systems, integrity testing, and related concepts, rather than standards to be contemplated during the planning/routing process.

PHMSA's "Safety Standards" Promulgated Pursuant to the Pipeline Safety Act

Another helpful tool in interpreting the meaning of the phrase "safety standards for interstate pipeline facilities or interstate pipeline transportation" is the extent to which the federal regulatory agency tasked with implementing those standards has asserted its authority. The Pipeline and Hazardous Materials Safety Administration (PHMSA) is a division of the U.S. Department of Transportation tasked with, among other things, the promulgation and enforcement of regulations issued pursuant to the terms of the Pipeline Safety Act. These regulations can be found at 49 C.F.R. Parts 190-195. The regulations are helpful in interpreting the Pipeline Safety Act because they offer insight into the PHMSA's perception of its authority under the act.

Multiple substantive matters are addressed in PHMSA's pipeline safety regulations, which are issued pursuant to its authority under the Pipeline Safety Act.[129] These requirements all appear to relate to the physical properties of the pipeline and associated facilities, the qualifications of the personnel employed in the construction, operation and maintenance of the pipelines, or emergency response requirements.

The regulations seem to recognize the different safety and environmental hazards presented by the siting of pipelines in different locations. For example, the regulations appear to establish varying standards for pipelines in rural areas, "unusually sensitive areas," and other geographical distinctions. It appears that these distinctions are made in order to establish different physical, emergency response and other requirements rather than to dictate a policy of siting pipelines away from these areas. State actors that infringe on this assertion of federal jurisdiction could be preempted by federal statute.

Thus, it appears that the PHMSA seems to view the extent of its authority over "safety standards for interstate pipeline facilities or interstate pipeline transportation" to extend to subjects such as the physical makeup up the pipeline, monitoring requirements, personnel qualifications, emergency response requirements, and related topics. If this is the extent of the reach of the Pipeline Safety Act's jurisdiction with respect to oil pipelines, state action would avoid preemption so long as it does not affect these aspects of oil pipeline operation.

Federal Jurisprudence Challenging State Action as Preempted by the Pipeline Safety Act

Finally, jurisprudence in the federal courts can assist in understanding the parameters of the federal authority to adopt "safety standards for interstate pipeline facilities or interstate pipeline transportation" and the express limitation on the states to adopt such standards. Although there is no case law evaluating a challenge to a state statute or regulatory action governing the location of an interstate pipeline, these cases help to draw the line between a permissible state activity governing pipeline activities and an impermissible foray into interstate pipeline safety standards expressly preempted by the Pipeline Safety Act.

In *Texas Midstream Gas Services v. City of Grand Prairie*,[130] a natural gas pipeline operator sought to enjoin the Grand Prairie City Council from enacting and enforcing a city ordinance known as "Section 10," requiring a special permit from the city in order to build a natural gas compressor station, and mandating that compressor stations comply with certain setback requirements, security requirements, aesthetic and noise level requirements.[131] Parties were also required to pave means of vehicular access to the facilities in order to obtain a permit.[132] In this instance, the pipeline operator had announced its intentions to build and operate the facility prior to the enactment of Section 10 by the Grand Prairie City Council.[133] The pipeline operator claimed, among other things, that Section 10 was preempted by the Pipeline

Safety Act. After a federal district court held that only that portion of Section 10 requiring a security fence was preempted by the Pipeline Safety Act and rejected the request for injunction regarding enforcement of the remainder of Section 10, the pipeline operator appealed to the U.S. Court of Appeals for the Fifth Circuit. That court affirmed the lower court's decision and rejected the request for an injunction, thus allowing the city to enforce Section 10 with the exception of the security fence requirement.[134]

With respect to the setback requirement, the court noted that

> The question is whether the setback requirement is a "safety standard." It is not. Along with the other provisions of Section 10, the setback requirement primarily ensures that bulky, unsightly, noisy compressor stations do not mar neighborhood aesthetics. City Council records reveal that Grand Prairie's primary motivation in adopting Section 10 was to preserve neighborhood visual cohesion, avoiding eyesores or diminished property values.[135]

The pipeline company claimed that, regardless of the purpose of the enactment, Section 10 had the effect of regulating fire safety, and thus was preempted by the Pipeline Safety Act.[136] The court rejected this argument, finding that "[a] local rule may incidentally affect safety, so long as the affect is not 'direct and substantial.'"[137] The court also addressed the scope of the PSHMA regulations as discussed above. In response to an assertion by the pipeline that Section 10 was preempted because the PHMSA regulations "address the location of compressor stations," the court clarified that "the PSA itself only preempts *safety* standards," and that the regulation cited by the pipeline "touches on compressor station location only as a means of effectuating this legislative directive. A regulation promulgated by an administrative agency cannot expand the unambiguously expressed preemptive scope set by Congress."[138] This decision helps illustrate the type of state or local actions related to pipeline facilities that the courts have held to be preempted by the Pipeline Safety Act's assertion of exclusive federal jurisdiction over "safety standards" for interstate pipeline facilities so long as the state or local action does not have "safety standards" for the pipeline as its primary intent.

Another U.S. Court of Appeals decision, *Kinley v. Iowa Utilities Board*,[139] illustrates the sort of state or local action that the courts have found would be preempted by the Pipeline Safety Act. In *Kinley*, the owner/operator of an interstate petroleum product pipeline challenged an Iowa state statute that established a "comprehensive state program supervising the intrastate and interstate transportation of solid, liquid or gaseous substances ... in order to protect the safety and welfare of the public."[140] The state statute, referred to as "Chapter 479" required that such facilities be permitted by the state.[141] The pipeline owner/operator challenged the state's assertion of jurisdiction over the operations of its interstate pipeline as preempted by the Pipeline Safety Act. After the district court agreed with the owner that the state's assertion of jurisdiction over interstate pipelines under Chapter 479 was preempted, the state appealed to the Eighth Circuit Court of Appeals. On appeal, the state acknowledged that certain aspects of Chapter 479 that explicitly dealt with safety were preempted by the Pipeline Safety Act. However, the state argued that there were certain "non-safety" provisions of Chapter 479, specifically hearing, permit and inspection provisions as well as financial responsibility requirements designed to protect the state's farmland and topsoil from damage due to the construction, operation and maintenance of the pipelines and to guarantee payment of property and environmental damages, that were not preempted.[142] The court disagreed, finding that the hearing, permit and inspection provisions of Chapter 479 are "so related to the federal safety regulations that they are preempted" by the Pipeline Safety Act, and further holding that "environmental and damage remedies provisions are not severable from the preempted hearing, permit and inspection provisions and thus are preempted as well."[143] The court thus affirmed the lower court's opinion and disallowed the application of Chapter 479 to the interstate pipeline.[144]

The example of *Kinley* is instructive to states seeking to regulate interstate pipelines. A state that is contemplating such action might analyze Iowa's Chapter 479 to avoid creating law that would, like Iowa's Chapter 479, be preempted by the Pipeline Safety Act. It should, however, be noted that with respect to the "environmental and damage remedies" provisions, the court found that these provisions were preempted by the Pipeline Safety Act not by virtue of their content but rather because they could not be severed from other aspects of Chapter 479. As the court noted in *Kinley*, the question of severability is not a constitutional question but instead is a matter of state law.[145]

Many states have statutes and regulations that govern interstate pipelines subject to the Pipeline Safety Act that appear not to have been deemed preempted by the Pipeline Safety Act. Although the lack of a successful preemption challenge does not conclusively demonstrate that the state statute would survive such a challenge, a state might nevertheless find that such statutes and regulations provide guidance in avoiding preemption.

Conflict Preemption of State Action Related to the Proposed Keystone XL Pipeline

If the state action is not expressly preempted, it may be preempted by what is commonly known as "conflict preemption." There appear to be two arguments that action regarding the siting of the proposed Keystone XL pipeline would be preempted by conflict with federal law and would therefore be unconstitutional. The first is that a siting statute would conflict with the president's authority to permit cross-border pipelines. It is difficult to comprehend how a state statute that addresses the location of a facility within its borders would "conflict" with an exercise of executive branch authority to authorize cross-border pipelines. There would likely be no operational conflict, as compliance with both the federal requirement and a state permitting/siting requirement is not only physically possible but has been achieved by numerous other international pipeline projects that have obtained presidential permits and also satisfied permitting and siting requirements of other states.[146] It is also not clear whether and how a state permitting/siting requirement would "stand as an obstacle to the accomplishment and execution of the full purposes and objectives of Congress." Although the State Department considered the environmental impacts of the entire Keystone XL project when conducting its NEPA review, the State Department's actual permitting authority extends only to that portion of the facility that crosses the U.S.-Canada border. According to its own terms, the purpose of the exercise of executive authority found in Executive Order 13337 is

> to expedite reviews of permits as necessary to accelerate the completion of energy production and transmission projects, and to provide a systematic method for evaluating and permitting the construction and maintenance of certain border crossings for land transportation .... that do not require the construction and maintenance of facilities connecting the United States with a foreign country, while maintaining safety, public health, and environmental protections.[147]

Viewed in isolation, some of this language could be interpreted to suggest a generic federal "purpose and objective" of accelerating completion of energy production and transmission projects, in which case state legislation which might impede the completion of such projects could be considered to be in conflict with the stated federal purpose. However, this appears likely to be too narrow a view of the objective of Executive Order 13337, which appears intended to expedite the processing of presidential permit applications. It therefore seems unlikely that a court would construe state action impacting the siting of an oil pipeline within the state's borders as "an obstacle to the accomplishment and execution of the full purposes and objectives" of the federal government as expressed in Executive Order 13337.

A second potential claim of "conflict preemption" would be an assertion that state action conflicts with the Pipeline Safety Act. It is impossible to draw a conclusion without knowing the nature of the state's action. Additionally, it is difficult to articulate an argument that a state statute related to oil pipeline siting would present either an "operational conflict" with the Pipeline Safety Act or would be deemed "an obstacle to the accomplishment and execution of the full purposes and objectives" of the Pipeline Safety Act, which are stated as "to provide adequate protection against risks to life and property posted by pipeline transportation and pipeline facilities by improving the regulatory and enforcement authority of the Secretary of Transportation."[148] So long as state action did not interfere with the accomplishment and execution of these objectives by the federal government and did not make it impossible for an entity to comply with its actions, the Pipeline Safety Act, and any regulations issued pursuant to the act, a court would likely not find that the state's action was invalidated by "conflict preemption" with the Pipeline Safety Act.

Field Preemption of State Action Related to the Proposed Keystone XL Pipeline

The third and final potential preemption of state action is what is commonly known as "field preemption"; that is, where it is inferred from congressional action that Congress intended to remove a state's regulatory authority over an entire subject. However, the Supreme Court has held that

> When Congress has considered the issue of pre-emption and has included in the enacted legislation a provision explicitly addressing that issue, and when that provision provides a "reliable indicium of congressional intent with respect to state authority," "there is no need to infer congressional intent to pre-empt state laws from the substantive provisions" of the legislation.[149]

Because Congress explicitly addressed the issue of preemption in the Pipeline Safety Act at 49 U.S.C. Section 60104, there is no need to explore the notion of "field preemption" with respect to that legislation.

JUDICIAL REVIEW OF THE NEPA PROCESS FOR PERMITTING UNDER EXECUTIVE ORDER 13337

The State Department reviews environmental impacts prior to making the national interest determination. However, the legal necessity of the State Department's review under NEPA for projects under Executive Order 13337 is uncertain. Compliance with NEPA in this instance could mean a project once immune from judicial oversight is subject to court review.

Executive Order 13337

As discussed above, Executive Order 13337 delegates the President's authority to the State Department to issue permits for certain energy projects that cross international borders. Executive Order 13337 requires that the State Department seek the views of certain Agency heads, including the Secretary of the Interior and the Administrator of the Environmental Protection Agency, prior to issuing the determination.[150] While the introductory language of the order indicates one purpose of the process is to maintain "environmental protections," the order does not explicitly require consideration of environmental impacts. It states only that the permit be in the national interest. Executive Order 13337 is a modification of Executive Order No. 11423, which predates NEPA,[151] and thus, does not reference a NEPA review to inform the national interest determination.

Keystone XL NEPA Background

For the Keystone XL project, an environmental impact statement (EIS) was prepared under NEPA.[152] On January 28, 2009, the Department of State announced receipt of the application and its intent to conduct a NEPA review:

> With respect to the application submitted by Keystone, the Department of State has concluded that the issuance of the Presidential permit would constitute a major Federal action that may have a significant impact upon the environment within the meaning of the National Environmental Policy Act (NEPA) of 1969. For this reason, Department of State intends to prepare an EIS

to address reasonably foreseeable impacts from the proposed action and alternatives.[153]

State Department NEPA Regulations

The State Department's announcement was consistent with its regulations. Under 22 C.F.R. Section 161.7(c), the State Department will prepare an environmental assessment[154] for certain projects, including issuing permits for construction of international pipelines as provided under Executive Order 11423.[155]

Pursuant to its regulations, the State Department evaluates the environmental impacts under NEPA prior to making a national interest determination for a permit. For example, it prepared an environmental assessment for an oil pipeline that crossed from Alberta, Canada, to Superior, Wisconsin.[156] It also prepared an EIS for the earlier Keystone oil pipeline from Canada.[157]

Judicial Review

Courts have considered whether any environmental review prepared by the State Department for projects seeking a permit under Executive Order 13337 is subject to judicial review. Only triallevel courts have heard the issue, and the decisions in those courts have been split. The U.S. District Court for the District of Columbia held that there was no judicial review of Executive Order 13337 actions.[158] A year later, the U.S. District Court for the District of Minnesota, which was not bound by that precedent, held that a State Department NEPA review prepared for a permit under Executive Order 13337 was a final agency action subject to review by a court.

In order to sue the federal government, the government must waive of sovereign immunity. For federal agency actions, that waiver is typically provided by the Administrative Procedure Act (APA): "Agency action made reviewable by statute and final agency action for which there is no other adequate remedy in a court are subject to judicial review."[159] Courts have consistently held that an EIS is a final agency action subject to review under the APA. However, under accepted precedent, no NEPA review is required for actions by a President because NEPA applies to "agencies of the Federal government,"[160] and not to presidential actions.[161] Thus, although it seems no NEPA review was required under Executive Order 13337, the State Department's NEPA review may be an agency action that is judicially reviewable. By conducting a NEPA review, the State Department may have opened that evaluation to court scrutiny.

District Court Holdings Related to Executive Order 13337

District courts disagree over whether an EIS would be an *agency* action under these circumstances. The district court of D.C. held that it is not, relying on the fact that the Executive Order is a delegation by the President of a constitutional authority.[162] The court in that case focused on the issuance of the permit and not on the EIS, itself. Since the delegation did not change the underlying authority for the action, just who issues the permit, according to the D.C. district court, any activity by the State Department pursuant to that Executive Order is a presidential action, and not an agency action.[163]

The U.S. District Court for the District of Minnesota disagreed with the reasoning of the D.C. district court and found that an EIS by the State Department for a cross-border pipeline permit was reviewable.[164] The court noted the State Department's *Federal Register* notice in which it stated that the pipeline permit was a "major federal action."[165] The court noted that although the permit pertained to a pipeline crossing an international border, that did not excuse the State Department from analyzing the environmental impacts of the entire route. More important, the court held that the fact that the *permit* for the international crossing was a presidential action did not "convert the State Department's preparation of the FEIS into a presidential action."[166]

To some extent, the two courts are arguing about two different things: the issuance of the permit, and the issuance of the EIS. However, only the District of Minnesota separates the two actions, noting that the issuance of the permit was not before the court,[167] and allowing judicial review of the EIS. Further, evidence appears to support the notion that an EIS is an agency action separate from the State Department's issuance of a presidential permit under Executive Order 13337. First, the State Department, under APA notice and comment rulemaking created a rule stating it would prepare an EA for pipelines under that Executive Order.[168] If the State Department failed to complete a NEPA review while that rule was still in effect,[169] it could be liable under the APA for acting contrary to its own regulations. Second, there is no directive in the Executive Order to perform any type of environmental review. A permit could be issued without any NEPA review, but occurs because of a State Department regulation requiring an EA, consistent with NEPA. Thus the NEPA review could be characterized as a congressional delegation of authority by statute (NEPA) rather than a presidential delegation of authority via Executive Order 13337. Additionally, the State Department's description of its process indicates that the national interest determination is separate from the NEPA review, lending further support to the argument that there are two actions: "Following the release of the Final EIS, a review period begins to determine if the proposed project is in the national interest. This broader evaluation of the application extends beyond environmental impact, taking into account economic, energy security, foreign policy, and other relevant issues."[170]

In reaching its conclusion that issuing a permit under Executive Order 13337 is judicially untouchable, the D.C. district court reviewed Supreme Court precedent and noted that no decision is directly on point.[171] The Supreme Court has considered presidential actions only in the context of those authorized by statute.[172] In those cases, *Franklin v. Massachusetts*[173] and *Dalton v. Specter*,[174] the Court reviewed challenges to the sufficiency of agency or commission reports submitted to the President before he acted. The decisions of the President in each case were held to be insulated from judicial review because, regardless of the documents' content, the Court held the President had ultimate discretion. The reports submitted to the President as part of each underlying statutory directive did not change the nature of the authority: regardless of the content of the reports, the decision was still made by the President. In the case of *Franklin*, the statute did not require the President to use the data in the report.[175] In *Dalton*, the Court held that despite the statute's requirement limiting the President to approving or rejecting the list in its entirety, the ultimate authority was still presidential: no action was final until the President submitted his certificate of approval to congress.[176]

Distinctions between the Supreme Court cases and an EIS for Executive Order 13337 may be made. In those court cases, the documents given to the President were required by an underlying statute and deemed part of the delegated authority. In cases under Executive Order 13337, there is no presidential directive suggesting or requiring an environmental review. The EIS was a discretionary action of the agency based on a statutory requirement that agencies act "to the fullest extent possible" to review agency actions.[177] To make that distinction more clear, if a Secretary submitted a "view" under Executive Order 13337 Section 1(b)(2) stating it was not in the national interest to issue the permit, but the State Department issued the permit anyway, under Supreme Court precedent neither the "view" nor the permit issuance appears to be reviewable. Under the holdings in *Franklin* and *Dalton*, that "view" would not be final and the permit issuance would be part of the delegated presidential authority because the "view" was advisory and part of the delegated presidential authority. In contrast, if the EIS finding was in error, well-established case law would allow challenge of the EIS as "arbitrary and capricious" under the APA.[178] While that might not change the State Department's determination of the project's being in the national interest, it would make the EIS a final agency action reviewable under accepted legal precedent.

Second, the reports in *Franklin* and *Dalton* were found not to be final actions because the presidential decision still had to be made. Whereas, courts have held repeatedly that a completed EIS is a final action.[179] In fact, in *Franklin*, the Court noted the ways that the census report was not an APA agency action: it was "not promulgated to the public in the Federal Register, no official administrative record is generated, and its effect on reapportionment is felt only after the President makes the necessary calculations and reports the results to Congress."[180] In contrast, each of those factors occurs for an Executive Order 13337 NEPA review: it was promulgated to the public, there is an administrative record, and under Supreme Court precedent, injury under NEPA occurs when the agency fails to comply with the law.[181] Because the State Department prepared a supplemental EIS to address EPA's concerns with the draft EIS, the State Department appears to concede that its EIS was different from the documents at issue in *Franklin* and *Dalton*.

The *Franklin* Court said that until the President acted there was "no determinate agency action to challenge, since the President, not the Secretary, takes the final action that affects the States."[182] In the case of Keystone XL, while issuing a permit is a final action taken under presidential authority, the EIS may also be a final action, which is taken under separate statutory authority. The D.C. court's holding did not recognize that distinction. Instead, the D.C. court found that the permit was a presidential action not subject to judicial review. Minnesota found that the EIS was a final agency action subject to judicial review.

End Notes

[1] This is an abbreviated discussion of the physical characteristics of the proposed Keystone XL pipeline for the purposes of introducing the legal issues related to the pipeline. More information is available at CRS Report R41668, *Keystone XL Pipeline Project: Key Issues*, by Paul W. Parfomak, et al.

[2] U.S. Department of State, *Supplemental Draft Environmental Impact Statement for the Keystone XL Oil Pipeline Project*, p. 1-4 (April 15, 2011).

[3] *See, e.g.*, S.D. Codified Laws §49-41B-4.1 (requiring a state permit and the approval of the state legislature prior to construction of a "trans-state" transmission facility, defined to include pipelines).

[4] Pipeline safety is governed by the federal Office of Pipeline Safety, an agency within the Department of Transportation, which exercises authority pursuant to the Hazardous Liquid Pipeline Safety Act of 1979, as amended (49 U.S.C. §60101 *et seq.*).

[5] The Federal Energy Regulatory Commission (FERC) regulates oil pipeline rates pursuant to the authority transferred to it by Title XVIII of the National Energy Policy Act of 1992 (P.L. 102-486, 42 U.S.C. §7172 note).

[6] The State Department was delegated authority to issue a permit for cross-boundary pipelines by the President under Executive Order 11423 of 1968, as revised in 2004 by Executive Order 13337.

[7] Exec. Order No. 13337, §1(g), *Issuance of Permits With Respect to Certain Energy-Related Facilities and Land Transportation Crossings on the International Boundaries of the United States*, 69 *Federal Register* 25299 (May 5, 2004).

[8] 42 U.S.C. §§4321 et seq. This report assumes some familiarity with the NEPA process. To learn details about that process, see CRS Report RS20621, *Overview of National Environmental Policy Act (NEPA) Requirements*, by Kristina Alexander, or see the discussion of NEPA in CRS Report R41668, *Keystone XL Pipeline Project: Key Issues*, by Paul W. Parfomak et al.

[9] 75 Fed. Reg. 20653 (April 20, 2010).

[10] Letter from Assistant Administrator, U.S. Environmental Protection Agency, to Assistant Secretary of State, U.S. Department of State (July 16, 2010) ("we think that the Draft EIS does not provide the scope or detail of analysis necessary to fully inform decision makers and the public, and recommend that additional information and analysis be provided"), available at http://yosemite.epa.gov/oeca/webeis.nsf/%28PDFView%29/20100126/$file/20100126.PDF.

[11] 76 Fed. Reg. 22744 (April 22, 2011).

[12] Letter from Assistant Administrator, EPA, to Assistant Secretaries of State, State Department (June 6, 2011) ("additional analysis is necessary... [including] improv[ing] the analysis of oil spill risks and alternative pipeline routes"), available at http://environblog.jenner.com/files/here-6.pdf.

[13] 76 Fed. Reg. 53525 (August 26, 2011).

[14] 40 C.F.R §1502.19 (circulation of EIS); 40 C.F.R §1505.2 (record of decision).

[15] P.L. 112-78.

[16] P.L. 112-78, § 501.

[17] P.L. 112-78, § 501(b)(3).

[18] Press Release, The White House, Office of the Press Secretary, Statement by the President on the Keystone XL Pipeline (January 18, 2012), http://www.whitehouse.gov/the-press-office/2012/01/18/statement-president-keystone-xlpipeline; Media Note, U.S. Department of State, Office of the Spokesperson, Denial of Keystone XL Pipeline Application (January 18, 2012), http://www.cq.com/pdf/govdoc-4012555.

[19] Media Note, U.S. Department of State, Office of the Spokesperson, Denial of Keystone XL Pipeline Application (January 18, 2012), http://www.cq.com/pdf/govdoc-4012555.

[20] *Id.*

[21] *Id.*

[22] Memorandum from President Barack Obama to the Secretary of State on Implementing Provisions of the Temporary Payroll Tax Cut Continuation Act of 2011 Relating to the Keystone XL Pipeline Permit, http://www.whitehouse.gov/the-press-office/2012/01/18/presidential-memorandum-implementing-provisions-temporary-payroll-tax-cu.

[23] *Id.*

[24] *See* United States v. Guy W. Capps, Inc., 204 F.2d 655 (4th Cir. 1953), *affirmed on other grounds by* 348 U.S. 296 (1955) ("We think that whatever the power of the executive with respect to making executive trade agreements regulating foreign commerce in the absence of action by Congress, it is clear that the executive may not through entering into such an agreement avoid complying with a regulation prescribed by Congress. Imports from a

foreign country are foreign commerce subject to regulation, so far as this country is concerned, by Congress alone.").

[25] Exec. Order No. 11423, *Providing for the performance of certain functions heretofore performed by the President with respect to certain facilities constructed and maintained on the borders of the United States*, 33 Fed. Reg. 11741 (August 16, 1968).

[26] Exec. Order No.13337, *Issuance of Permits With Respect to Certain Energy-Related Facilities and Land Transportation Crossings on the International Boundaries of the United States*, 69 Fed. Reg. 25299 (May 5, 2004).

[27] Exec. Order No. 11423, 33 Fed. Reg. at 11741.

[28] Exec. Order No. 13337, 69 Fed. Reg. at 25299.

[29] *Id.* at 25230.

[30] 33 Fed. Reg. at 11741.

[31] 69 Fed. Reg. at 25299.

[32] 659 F. Supp. 2d 1071, 1078 (D.S.D. 2009). This Keystone pipeline project preceded the Keystone XL pipeline.

[33] *Id.* at 1078, 1078 n.5.

[34] *Id.* at 1081-82.

[35] *Id.* at 1081.

[36] *Id.*

[37] 689 F. Supp. 2d 1147 (D. Minn. 2010).

[38] *Id.* at 1162.

[39] *Id.*

[40] *Id.* at 1163 (citing 38 U.S. Att'y Gen. 163 (1935); 30 U.S. Op. Att'y Gen. 217 (1913); 24 U.S. Op. Att'y Gen. 100 (1902); 22 Op. Att'y Gen. 408 (1899); and Natural Resources Defense Council (NRDC) v. U.S. Department of State, 658 F. Supp. 2d 105, 109 (D.D.C. 2009)).

[41] *Id.*

[42] Robert J. Delahunt, *Federalism Beyond the Water's Edge: State Procurement Sanctions and Foreign Affairs*, 37 STAN. J. INT'L L. 1, 25 (2001).

[43] 22 Op. Att'y Gen. 13 (1898).

[44] *Id.* (citing Senate Doc. 122, at 70).

[45] *Id.* (citing Senate Doc. 122, at 70).

[46] *Id.*

[47] *Id.*

[48] *Id.* (reciting from a May 11, 1897 letter from the State Department).

[49] *Id.* (reciting from a May 11, 1897 letter from the State Department).

[50] Memorandum in Support of Defendant's Motion to Dismiss Plaintiff's First Amended Complaint, at 12 (October 20, 2008), Natural Resources Defense Council, Inc. v. United States Department of State, 658 F. Supp. 2d 105 (D.D.C. 2009) (No. 08-CV-01363).

[51] Foreign Cables, 22 Op. Att'y Gen. 13 (January 18, 1898).

[52] *Id.*

[53] *Id.*

[54] *Id.* (citing Forty-ninth Congress, second session, Senate Doc. 122, at 63).

[55] *Id.* (citing Senate Doc. 122, at 76).

[56] *Id.*

[57] *Id.*

[58] *Id.*

[59] Cuba-Cables, 22 Op. Att'y Gen. 408 (1899).

[60] *Id.*

[61] *Id.*

[62] *Id.*

[63] *Id.*

[64] Diversion of Water from Niagara River, 30 Op. Att'y Gen. 217 (1913).

[65] *Id.*

[66] *Id.*
[67] *Id.*
[68] *Id.*
[69] *Id.*
[70] *Id.*
[71] Exec. Order No. 11423, 33 Fed. Reg. 11741 (August 16, 1968).
[72] *See* Exec. Order No. 13337, § 2(a), 69 Fed. Reg. 25299, 25300-01 (May 5, 2004).
[73] P.L. 67-8; P.L. 92-434. *See* Memorandum in Support of Defendant's Motion to Dismiss Plaintiff's First Amended Complaint, at 13 n.5 (October 20, 2008), Natural Resources Defense Council, Inc. v. United States Department of State, 658 F. Supp. 2d 105 (D.D.C. 2009) (No. 08-CV-01363).
[74] P.L. 96-72. This act prohibited the export from the United States or any of its territories or possessions of domestically produced crude oil transported by pipeline over rights-of-way granted under the Trans-Alaska Pipeline Authorization Act, P.L. 93-153.
[75] P.L. 66-146. This act restricted the export of crude oil transported by pipelines on federal lands, with an exemption for certain crude oil. This act allowed the export of unexempted oil provided the President made certain findings, recommended the export, and reported to Congress, and Congress enacted a joint resolution approving such export.
[76] P.L. 94-258. This act provided that before certain exports of petroleum could occur, the President must make, publish, and report to Congress the express finding that they will not diminish the total quantity or quality of petroleum available to the United States, are in the national interest, and are in accord with the Export Administration Act.
[77] P.L. 95-372. This act provided that before certain oil could be exported, the President must make and publish an express finding that the exports will not increase reliance on imported oil, are in the national interest, and are in accord with the Export Administration Act.
[78] P.L. 94-163. This act required the President to issue a rule prohibiting the export of crude oil produced in the United States, except that he may exempt from the prohibition crude oil exports that he determines to be consistent with the national interest and with statutory policies regarding the availability of domestic supplies.
[79] P.L. 99-440. This act prohibited exports of crude oil and refined petroleum products to South Africa, and the prohibition was lifted by Executive Order 12769, which was allowed under the act.
[80] *See, e.g.*, Buttfield v. Stranahan, 192 U.S. 470, 492-93 (1904)("The power to regulate commerce with foreign nations is expressly conferred upon Congress, and being an enumerated power is complete in itself, acknowledging no limitations other than those prescribed in the Constitution. Lottery Case, 188 U.S. 321, 353-356; Leisy v. Hardin, 135 U.S. 100, 108."); Japan Line, Ltd. v. County of Los Angeles, 441 U.S. 434, 449 n.13 (1979) (citing *Buttfield* for the proposition that Congress's power over foreign commerce is "exclusive and absolute"); California Bankers Ass'n v. Shultz, 416 U.S. 21 (1974).
[81] Board of Trustees of University of Illinois v. United States, 289 U.S. 48, 56 (1933). The Court stated that "[a]s an exclusive power, its exercise may not be limited, qualified or impeded to any extent by state action." *Id.* at 56-57. The Court also stated that "[i]n international relations and with respect to foreign intercourse and trade the people of the United States act through a single government with unified and adequate national power." *Id.* at 59.
[82] 22 U.S. 1, 193-94 (1824).
[83] United States v. Clark, 435 F.3d 1100, 1113 (9th Cir. 2006), *cert. denied*, 549 U.S. 1343 (2007).
[84] 689 F. Supp. 2d 1147 (D. Minn. 2010). This case is discussed in greater detail above at p. 5-6.
[85] *Id.* at 1162.
[86] *Id.* at 1163.
[87] 659 F. Supp. 2d 1071 (D. S.D. 2009).
[88] *Id.* at 1081 (emphasis added).
[89] Youngstown Sheet and Tube Co. v. Sawyer, 343 U.S. 579, 637 (1952) (Jackson, J. concurring).

[90] *See* United States v. Morrison, 529 U.S. 598, 607 (2000) ("Due respect for the decisions of a coordinate branch of Government demands that we invalidate a congressional enactment only upon a plain showing that Congress exceeded its constitutional bounds.").

[91] Neb. Laws LB1 (November 22, 2011), available at http://nebraskalegislature.gov/FloorDocs/ Current/PDF/Final/ LB1_S1.pdf.

[92] Neb. Laws. LB4 (November 22, 2011), available at http://nebraskalegislature.gov/FloorDocs/ Current/PDF/Final/ LB4_S1.pdf.

[93] U.S. Const., Art. I, §8, cl. 3.

[94] *See generally*, Johnny Killian, George Costello & Kenneth Thomas, *Constitution of the United States of America: Analysis and Interpretation* 223-227 (2002).

[95] *Id.* at 221-222.

[96] Gibbons v. Ogden, 22 U.S. (9 Wheat.) 1 (1824); Wilson v. Black-Bird Creek Marsh Co., 27 U.S. (2 Pet.) 245 (1829); Cooley v. Board of Wardens of Port of Philadelphia, 53 U.S. (12 How.) 299 (1851).

[97] "When a state statute clearly discriminates against interstate commerce, it will be struck down ... unless the discrimination is demonstrably justified by a valid factor unrelated to economic protectionism." Wyoming v. Oklahoma, 502 U.S. 437, 454 (1992) *quoting* City of Philadelphia v. New Jersey, 437 U.S. 617, 624 (1978)). *See* Maine v. Taylor, 477 U.S. 131 (1986) (upholding state ban on importation of out-of-state baitfish based on possible presence of parasites and nonnative species in baitfish shipments).

[98] Discriminatory laws motivated by "simple economic protectionism" are subject to a "virtually per se rule of invalidity," City of Philadelphia v. New Jersey, 437 U.S. 617 (1978), Dean Milk Co. v. City of Madison, Wisconsin, 340 U.S. 349 (1951), Hunt v. Washington State Apple Advertising Comm., 432 U.S. 333 (1977), which can only be overcome by a showing that the State has no other means to advance a legitimate local purpose. Maine v. Taylor, U.S. 131 (1986). *See also* Brown-Forman Distillers v. New York State Liquor Authority, 476 U.S. 573 (1986).

[99] The most commonly cited case for this balancing test is *Pike v. Bruce Church, Inc*, 397 U.S. 137 (1970).

[100] *See, e.g.*, S.D. Codified Laws §49-41B-4.1 (requiring a state permit and the approval of the state legislature prior to construction of a "trans-state" transmission facility, defined to include pipelines). To CRS's knowledge, no such statute has ever been the subject of a legal challenge based on the dormant Commerce Clause.

[101] Energy Policy Act of 2005, P.L. 109-58, at §1221.

[102] Section 704(a) of the Telecommunications Act of 1996 (47 U.S.C. §332(c)(7)) allows state and local governments to make decisions regarding the siting of wireless communications facilities, subject to certain limitations, such as mandating that the local decision must not "prohibit or have the effect of prohibiting the provision of personal wireless services."

[103] Under the Federal Aid Highway Act (23 U.S.C. §101 *et seq.*), siting decisions for interstate highways are made solely by the states.

[104] 15 U.S.C. §717f(c).

[105] W. Lynn Creamery, Inc. v. Healy, 512 U.S. 186, 201 (1994).

[106] For example, an argument has been made that legislation that impacts only pipes in excess of a certain diameter, as has been proposed in draft legislation in Nebraska, might only affect interstate pipelines. If the law does not discriminate between in-state and out-of-state ownership, control, or other economic impacts of these pipelines, this distinction alone would not be grounds for a finding of discriminatory state action in violation of the dormant Commerce Clause.

[107] Minnesota v. Clover Leaf Creamery Co., 449 U.S. 456, 471 (1981) , *quoting Pike,* 397 U.S. at 142.

[108] *See* James D. Fox, *State Benefits Under the Pike Balancing Test of the Dormant Commerce Clause*, 1 Ave Maria L. Rev. 175 (20030.

[109] South-Central Timber Development, Inc. v. Wunnicke, 467 U.S. 82, 100 (1984).

[110] Note that the existence of a legitimate state interest would not override an express federal preemption of state authority. This topic will be treated later in this report.

[111] But see Lakehead Pipeline Company v. Illinois Commerce Commission, 696 N.E.2d 345 (Ill. App. 3rd Dist. 1998).

[112] Dakota, Minnesota & Eastern Railroad Corp., v. South Dakota, 236 F. Supp. 2d 989, 1018 (S.D.S.D. 2002)(finding dormant Commerce Clause violation because administrative burdens imposed on a pipeline project would prevent the pipeline from being built).

[113] P.L. 90-481.

[114] P.L. 96-129.

[115] P.L. 107-355.

[116] U.S. Const. Art. VI, cl. 2.

[117] Gibbons v. Ogden, 22 U.S. (9 Wheat.) 1, 211 (1824). *See also* McCulloch v. Maryland, 17 U.S. (4 Wheat.) 316, 436 (1819).

[118] California Federal Savings and Loan v. Guerra, 479 U.S. 272, 280 (1987), *citing* Jones v Rath Packing Co., 430 U.S. 519, 525 (1977).

[119] *Id.* at 281.

[120] *Id., quoting* Florida Lime & Avocado Growers, Inc. v. Paul, 373 U.S. 132, 142-143 (1963).

[121] *Id., quoting* Hines v. Davidowitz, 312 U.S. 52, 67 (1941).

[122] *Id.* at 280-281, *quoting* Rice v. Santa Fe Elevator Corp., 331 U.S. 218, 230 (1947).

[123] 49 U.S.C. §60104(c) (emphasis added).

[124] 49 U.S.C. §60104(e)

[125] 49 U.S.C. §60104(c).

[126] Presumably safety concerns often are a factor in pipeline siting.

[127] The standards are referred to as "government-cited" because the Pipelines and Hazardous Materials Safety Administration incorporates by reference into its regulations some of these industry standards. The standards are available at http://publications.api.org/ (registration is required).

[128] *Id.*

[129] See 49 C.F.R. parts 190–195.

[130] 608 F.3d 200 (3rd Cir. 2010).

[131] *Id.* at 203.

[132] *Id.* at 203-204.

[133] *Id.* at 203.

[134] *Id.* at 212.

[135] *Id.* at 211.

[136] *Id.*

[137] *Id., quoting* English v. Gen. Elec. Co., 496 U.S. 72, 85 (1990).

[138] *Id., citing* Cipollone v. Liggett Group, Inc., 505 U.S. 504. 517 (1992) (emphasis in original).

[139] 999 F.2d 354 (8th Cir. 1993).

[140] *Id.* at 356.

[141] *Id.*

[142] *Id.* at 357.

[143] *Id.* at 360.

[144] *Id.* The analysis in the *Kinley* decision is brief, in part because the court found that many of the issues raised had already been adjudicated in *ANR Pipeline Co. v. Iowa State Commerce Commission*, 828 F.2d 465 (8th Cir 1987). In that case, the U.S. Court of Appeals heard a similar preemption challenge to a previous iteration of Iowa Code Chapter 479 brought by a pipeline company that had been fined by the state due to alleged violations of the law. The more expansive discussion of the previous iteration of the preempted Iowa law found in that decision may also be instructive is seeking to avoid state action preempted by the Pipeline Safety Act.

[145] *Id.* at 359.

[146] *See, e.g.*, the "Alberta Clipper" pipeline project, the border-crossing of which was authorized by the U.S. Department of State in CITE. Enbridge Pipelines LLC has also obtained permits for the siting of that portion of the Alberta Clipper facility that is located in the state of North Dakota., pursuant to that state's statutory and regulatory requirements.

[147] *Executive Order 13337—Issuance of Permits With Respect to Certain Energy-Related Facilities and Land Transportation Crossings on the International Boundaries of the United States*, 69 Fed. Reg. 25229 (April 30, 2004).

[148] 49 U.S.C. §60102(a).

[149] Cipollone v. Liggett Group, Inc, 505 U.S. 504, 517 (1992), *quoting* Malone v. White Motor Corp, 435 U.S.497, 505 (1978) *and* California Federal Savings and Loan Assn v. Guerra, 479 U.S. 272, 282 (1987).

[150] Exec. Order No. 13337 §1(b). 69 Fed. Reg. 25299 (May 5, 2004).

[151] Executive Order No. 11423 was issued in 1968 (33 Fed. Reg. 11741 (August 20, 1968)), whereas NEPA was enacted on January 1, 1970.

[152] See Keystone XL report for details on the Keystone NEPA process.

[153] 74 Fed. Reg. 5019, 5020 (January 28, 2009).

[154] Under NEPA, an environmental assessment (EA) is a review to see whether an agency action might result in significant environmental impacts. If an agency foresees significant impacts, it may prepare an environmental impact statement (EIS) directly, rather than first preparing an EA.

[155] 45 Fed. Reg. 59553 (September 10, 1980). The State Department's NEPA regulations establish categories of action subject to NEPA and the environmental review procedures applicable to those actions. For actions determined to have a "significant" effect on the environment, an EIS must be prepared. 22 C.F.R. §161.7(a). An environmental assessment (EA) may be prepared to make that determination. 22 C.F.R. §161.7(c)(1). An EA is generally not necessary if an action's environmental impacts are known to be significant or if the action normally requires an EIS (§161.8(b)).

[156] *See* Sierra Club v. Clinton, 689 F. Supp. 2d 1147 (D. Minn. 2010).

[157] *See* NRDC v. U.S. Department of State, 658 F. Supp. 2d 105 (D.D.C. 2009).

[158] NRDC v. U.S. Department of State, 658 F. Supp. 2d 105 (D.D.C. 2009).

[159] 5 U.S.C. §704.

[160] 42 U.S.C. §4332(2)(C).

[161] *See* Alaska v. Carter, 462 F. Supp. 1155 (D. Alaska 1978); Utah Association of Counties v. Bush, 316 F. Supp. 2d 1172 (D. Utah 2004).

[162] NRDC v. State, 658 F. Supp. 2d 105 (D.D.C. 2009).

[163] *Id.* at 109 (D.D.C. 2009).

[164] Sierra Club v. Clinton, 689 F. Supp. 2d 1147, 1157 fn3 (D. Minn. 2010) ("the Court respectfully disagrees with th[at] decision[] insofar as [it] hold[s] that any action taken by the State Department pursuant to an executive order, and in particular the preparation of an EIS for a major federal action, is not subject to judicial review under the APA").

[165] *Id.* at 1157.

[166] *Id.* at 1157.

[167] The issuance of the permit was not before the court. *Id.* at 1161.

[168] *See* 45 Fed. Reg. 59553 (September 10, 1980).

[169] The State Department could remove that regulation pursuant to a formal rulemaking procedure and may then not be bound by NEPA.

[170] See, State Department Fact Sheet, p. 2 (August 26, 2011), available at http://www.Keystonepipeline-xl.state.gov/clientsite/keystonexl.nsf?Open.

[171] Similarly, the case Utah Ass'n of Counties v. Bush, 316 F. Supp. 2d 1172 (D. Utah, 2004), can be distinguished as the EIS in that case was prepared in support of an agency recommendation (as opposed to an agency decision) and was deemed advisory, and not final. Additionally, in *Tulare County v. Bush*, 185 F. Supp. 2d 18 (D.D.C. 2001), cited by the court, the challenge was based on a failure to prepare an EIS.

[172] As noted above, Executive Order 13337 is derived from constitutional authority.

[173] Franklin v. Massachusetts, 505 U.S. 788 (1992). This case challenged the Secretary of Commerce census report methodology as authorized under 13 U.S.C. §141. The Supreme Court described the process as having three parts. First, the Secretary of Commerce counts the U.S. population. Second, Commerce reports that total population to the President. And third, the President transmits a statement to Congress of the number of representatives for each state.

[174] Dalton v. Specter, 511 U.S. 462 (1994). This case challenged base closures in which a commission submits recommended closures to the President who certifies approval of the recommendations to Congress.

[175] Franklin, 505 U.S. at 797.

[176] Dalton, 511 U.S at 469.

[177] 42 U.S.C. §4332.

[178] Kleppe v. Sierra Club, 427 U.S. 290 (1976).

[179] See Ohio Forestry Association v. Sierra Club, 523 U.S. 726, 737 (1998) ("a person with standing who is injured by a failure to comply with the NEPA procedure may complain of that failure at the time the failure takes place, for the claim can never get riper").

[180] Franklin, 505 U.S. at 796.

[181] Ohio Forestry Association v. Sierra Club, 523 U.S. 726 (1998).

[182] Franklin, 505 U.S. at 799.

In: The Keystone XL Pipeline Project
Editors: G. B. Flynn and D. Burnett
ISBN: 978-1- 62081-221-1

Chapter 3

FINAL ENVIRONMENTAL IMPACT STATEMENT FOR THE PROPOSED KEYSTONE XL PROJECT*

United States Department of State

INTRODUCTION

In September 2008, TransCanada Keystone Pipeline, LP (Keystone) filed an application for a Presidential Permit with the U.S. Department of State (DOS) to build and operate the Keystone XL Project. The proposed Project would have the capacity to transport 700,000 barrels per day (bpd) of crude oil to delivery points in Oklahoma and southeastern Texas.

This Executive Summary of the final environmental impact statement (final EIS) summarizes the proposed Project, including the purpose of and need for the Project, and the major conclusions and areas of concern raised by agencies and the public. More detailed information on the proposed Project, alternatives to the proposed Project, and the associated potential environmental impacts is presented in the final EIS that is provided in the CD in the sleeve on the back page.

---

* This is an edited, reformatted and augmented version of the Executive Summary of United States Department of State Bureau of Oceans and International Environmental and Scientific Affairs, dated August 26, 2011.

PRESIDENTIAL PERMITTING PROCESS

All facilities which cross the international borders of the United States require a Presidential Permit. For liquid hydrocarbon pipelines, the President, through Executive Order 13337, directs the Secretary of State to decide whether a project is in the national interest before granting a Presidential Permit.

As part of the Presidential Permit review process, DOS determined that it should prepare an EIS consistent with the National Environmental Policy Act (NEPA). DOS is the lead federal agency for the NEPA environmental review of the Proposed Project because the need for a Presidential Permit is the most substantial federal decision related to the Proposed Project. To assist in preparing the EIS, DOS retained an environmental consulting firm, Cardno ENTRIX, following DOS guidelines on third-party contracts. The DOS environmental and safety review of the proposed Project that lead to the final EIS was conducted for nearly 3 years and included consultations with the third-party contractor, cooperating agencies, and scientists and engineers with expertise in key areas of concern related to the proposed Project.

ABOVEGROUND FACILITIES

- 30 pump stations on 5- to15-acre sites
- Delivery facilities at Cushing, Oklahoma and Nederland and Moore Junction, Texas
- Densitometer sites located at all injection points and at all delivery points
- 112 mainline valves along pipeline and 2 mainline valves at each pump station
- Tank farm at Cushing, Oklahomaon a 74-acre site

The determination of national interest involves consideration of many factors, including energy security; environmental, cultural, and economic impacts; foreign policy; and compliance with relevant federal regulations. Before making a decision, DOS will consult with the eight federal agencies identified in Executive Order 13337: the Departments of Energy, Defense, Transportation, Homeland Security, Justice, Interior, and Commerce, and the Environmental Protection Agency (EPA). DOS will also solicit public input on the national interest determination by accepting written comments and holding comment meetings in the six states traversed by the proposed route and in Washington, D.C.

Figure ES-1 lists the major events, public outreach activities, and other details of the environmental review and national interest determination processes.

SUMMARY OF THE KEYSTONE XL PROJECT

The proposed Keystone XL Project consists of a crude oil pipeline and related facilities that would primarily be used to transport Western Canadian Sedimentary Basin crude oil from an oil supply hub near Hardisty, Alberta Canada to delivery points in Oklahoma and Texas. The proposed Project would also be capable of transporting U.S. crude oil to those delivery points. The U.S. portion of the pipeline would begin near Morgan, Montana at the international border of the United States and extend to delivery points in Nederland and Moore Junction, Texas. There would also be a delivery point at Cushing, Oklahoma. These three delivery points would provide access to many other U.S. pipeline systems and terminals, including pipelines to refineries in the U.S. Gulf Coast region. Market conditions, not the operator of the pipeline, would determine the refining locations of the crude oil.

The proposed Keystone XL pipeline would consist of approximately 1,711 miles of new 36-inch-diameter pipeline, with approximately 327 miles of pipeline in Canada and 1,384 miles in the U.S. Figure ES-3 depicts the three segments of the proposed Project in the U.S. As noted in that illustration, the proposed Project would connect to the northern and southern ends of the existing Cushing Extension of the Keystone Oil Pipeline System.

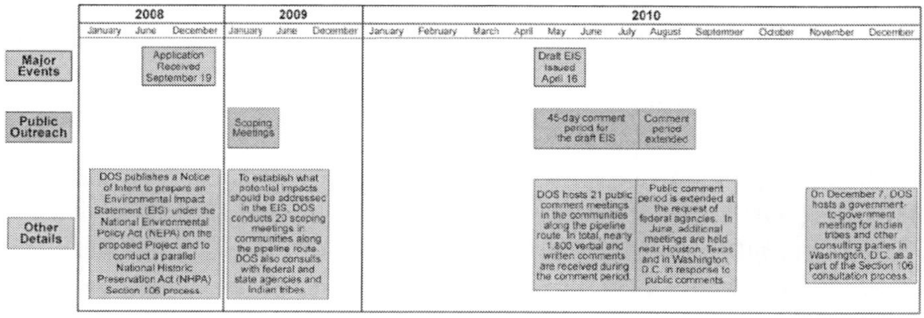

Figure ES-1. (Continued)

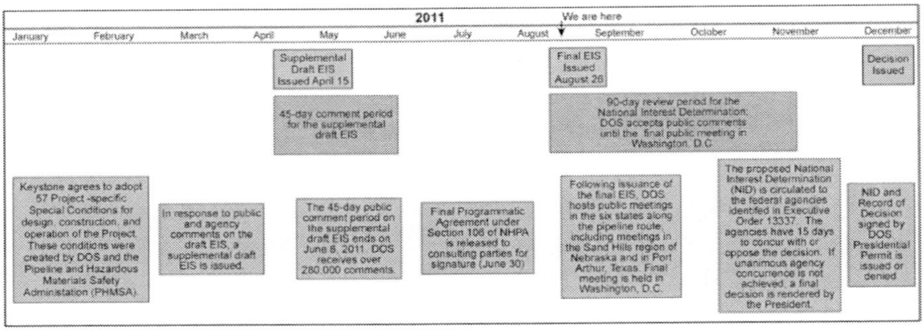

Figure ES-1. U.S. Department of State Environmental and National Interest Determination Review Processes.

Figure ES-2. 36-Inch-Diameter Crude Oil Pipe.

Figure ES-4 illustrates the construction sequence that would be followed for the proposed Project. The proposed Project would also include 30 electrically operated pump stations, 112 mainline valves, 50 permanent access roads, and a new oil storage facility in Cushing, Oklahoma. If market conditions change, the capacity of the proposed Project could be increased to 830,000 bpd by increasing pumping capacity at the proposed pump stations.

The overall proposed Keystone XL Project is estimated to cost $7 billion. If permitted, it would begin operation in 2013, with the actual date dependant on the necessary permits, approvals, and authorizations.

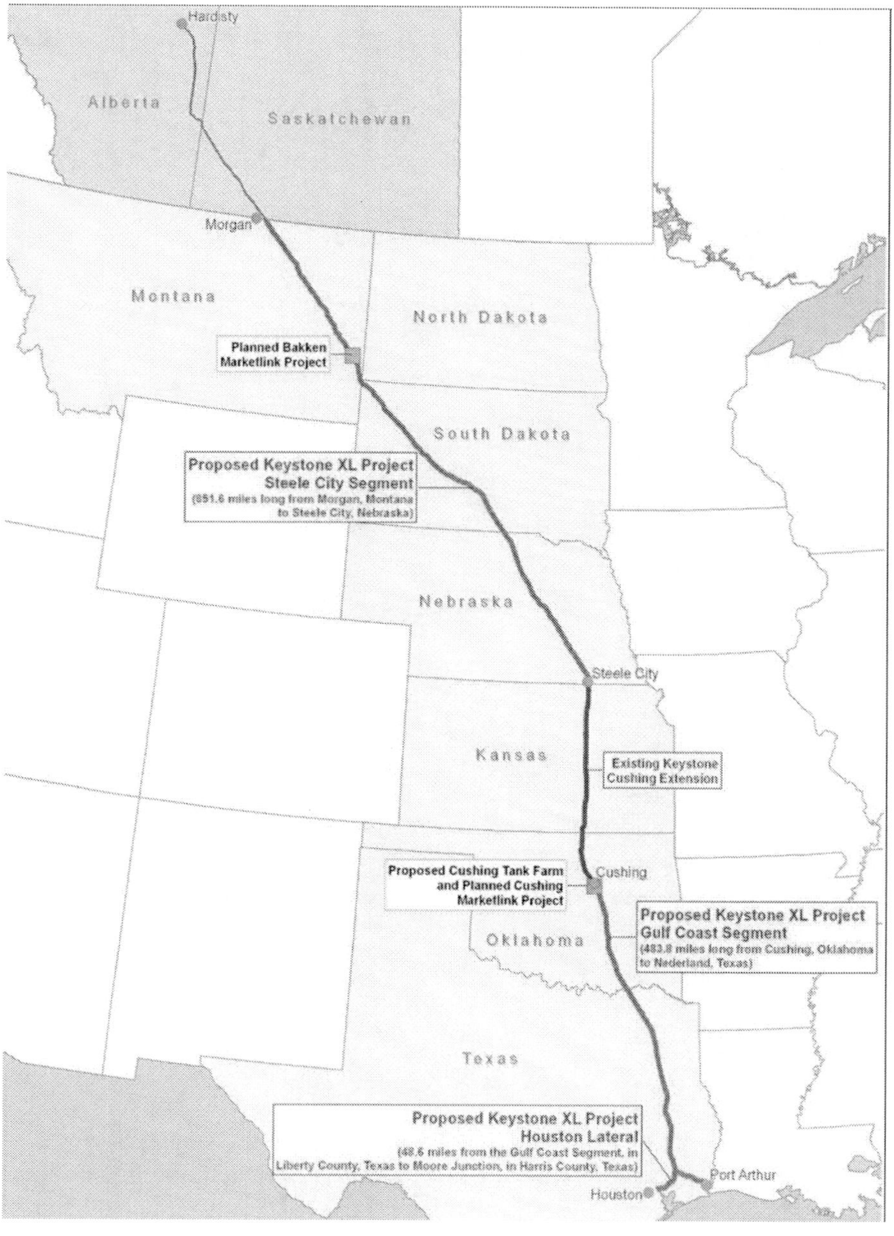

Figure ES-3. Proposed Pipeline Route.

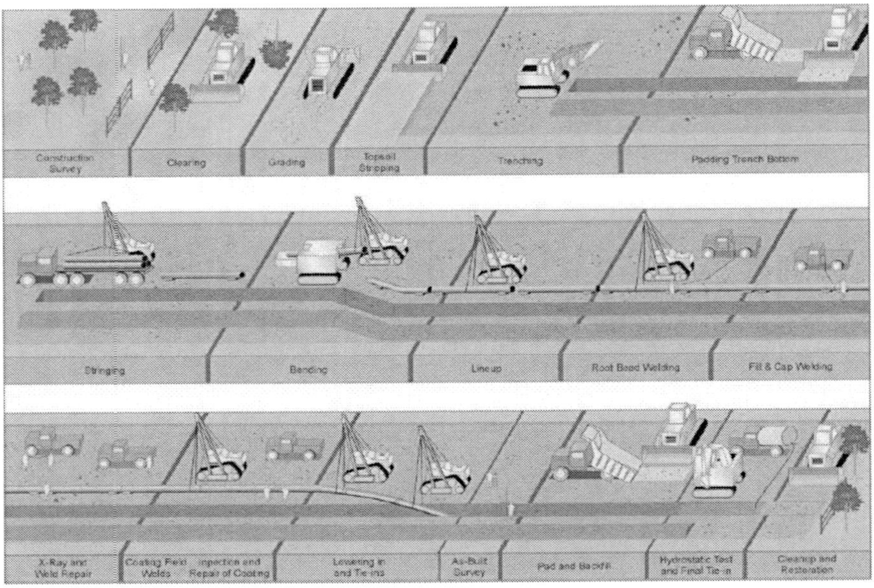

Figure ES-4. Typical Pipeline Construction Sequence.

Transport of Canadian Oil Sands Crude Oil

The proposed Keystone XL Project would primarily transport crude oil extracted from the oil sands areas in Alberta, Canada. Oil sands (which are also referred to as tar sands) are a combination of clay, sand, water, and bitumen, which is a material similar to soft asphalt. Bitumen is extracted from the ground by mining or by injecting steam underground to heat the bitumen to a point where it liquefies and can be pumped to the surface.

Bitumen is treated in several ways to create crude oil suitable for transport by pipeline and refining. The types of Canadian crude oil that would be transported by the proposed Project would primarily consist of synthetic crude oil and diluted bitumen.

Synthetic crude oil is produced from bitumen using refining methods – a process termed upgrading – that in general converts bitumen into lighter liquid hydrocarbons. In other words, the bitumen is converted into a crude oil similar to conventional crude oil.

Diluted bitumen – often termed dilbit – consists of bitumen mixed with a diluent, which is a light hydrocarbon liquid such as natural gas condensate or refinery naphtha. The bitumen is diluted to reduce its viscosity so that it is in a more liquid form that can be transported via pipeline. Dilbit is also processed to remove sand, water, and other impurities. The diluents in dilbit are integrally combined with the bitumen to form a crude oil that is a homogenous mixture that does not physically separate when released.

Both synthetic crude oil and dilbit are similar in composition and quality to the crude oils currently transported in pipelines in the U.S. and being refined in Gulf Coast refineries. Neither type of crude oil requires heating for transport in pipelines.

PIPE SPECIFICATIONS

- Material: High-strength X70 steel pipe, API 5L
- Outside diameter: 36 inches
- Operating Pressure: 1,308 psig
- External Coating: fusion-bonded epoxy

Transport of U.S. Crude Oil

In late 2010, Keystone Marketlink, LLC announced plans for two separate projects that would enable crude oil from domestic sources to be transported in the proposed Keystone XL Project. Those two projects, the Bakken Marketlink Project and the Cushing Marketlink Project, are considered "connected actions" under NEPA. The Bakken Marketlink Project would allow transport of up to 100,000 bpd of crude oil from the Bakken formation in the Williston Basin in Montana and North Dakota.

These fields have experienced high growth in the last few years as new technology has allowed the oil to be profitably extracted. Keystone currently has longterm commitments for transporting 65,000 bpd of crude oil in the proposed Keystone XL Project from the Bakken Marketlink Project.

The Cushing Marketlink Project would allow transport of up to 150,000 bpd on the proposed Project from the Cushing, Oklahoma area to the proposed Keystone XL Project delivery points in Texas.

Other Connected Actions

In addition to the Marketlink projects, there are two other types of connected actions associated with the Project: electrical distribution lines and substations that would provide power for the pump stations, and an electrical transmission line that would be required to ensure transmission system reliability when the proposed Project is operating at maximum capacity. Those projects would not be built or operated by Keystone, and the permit applications for those projects would be reviewed and acted on by other agencies. Although only limited information was available on the design, construction, and operation of the projects, DOS assessed the potential impacts of the projects based on currently available information.

PURPOSE OF AND NEED FOR THE KEYSTONE XL PROJECT

The primary purpose of the proposed Project is to provide the infrastructure necessary to transport Western Canadian Sedimentary Basin heavy crude oil from the U.S. border with Canada to delivery points in Texas in response to the market demand of Gulf Coast refineries for heavy crude oil. This market demand is driven by the need of the refiners to replace declining feed stocks of heavy crude oil obtained from other foreign sources with crude oil from a more stable and reliable source. Keystone currently has firm, long-term contracts to transport 380,000 bpd of Canadian crude oil to the Texas delivery points.

An additional purpose of the proposed Project is to transport Canadian heavy crude oil to the proposed Cushing tank farm in response to the market demand of refineries in the central and Midwest U.S. for heavy crude oil. Keystone also has firm contracts to transport 155,000 bpd of Canadian crude oil to Cushing, Oklahoma in the existing Keystone Oil Pipeline Project. If the proposed Project is approved and implemented, Keystone would transfer shipment of crude oil under those contracts to the proposed Project. Although there is sufficient pipeline capacity from Canada to the U.S. in general to accommodate projected additional imports of Canadian crude in the short to medium term, there is extremely limited pipeline transport capacity to move such crude oils to Gulf Coast refineries.

The 58 refineries in the Gulf Coast District provide a total refining capacity of approximately 8.4 million bpd, or nearly half of U.S. refining capacity. These refineries provide substantial volumes of refined petroleum product, such as gasoline and jet fuel, via pipeline to the Gulf Coast region as well as the East Coast and the Midwest.

In 2009, Gulf Coast refineries imported approximately 5.1 million bpd of crude oil from more than 40 countries. The top four suppliers were Mexico, Venezuela, Saudi Arabia, and Nigeria. Of the total volume imported, approximately 2.9 million bpd was heavy crude oil similar to the crude oil that would be transported by the proposed Project; Mexico and Venezuela were the major suppliers. However, imports of heavy crude oil from these two countries have been in steady decline while Gulf Coast refining capacity is projected to grow by at least 500,000 bpd by 2020, with or without the proposed Project.

PROJECT DESIGN AND SAFETY

The Pipeline and Hazardous Materials Safety Administration (PHMSA), a federal agency within the U.S. Department of Transportation, is the primary federal regulatory agency responsible for ensuring the safety of America's energy pipelines, including crude oil pipeline systems. As a part of that responsibility, PHMSA established regulatory requirements for the construction, operation, maintenance, monitoring, inspection, and repair of hazardous liquid pipeline systems.

In 2009, Keystone applied to PHMSA for a Special Permit to operate the proposed Project at a slightly higher pressure than allowed under the existing regulations. DOS worked with PHMSA to develop Project-specific Special Conditions that would have been incorporated into the Special Permit. However, in August 2010, Keystone withdrew its application to PHMSA for a Special Permit. However, to enhance the overall safety of the proposed Project, DOS and PHMSA continued working on Special Conditions specific to the proposed Project and ultimately established 57 Project-specific Special Conditions. As a result, Keystone agreed to design, construct, operate, maintain, and monitor the proposed Project in accordance with the more stringent 57 Project specific Special Conditions in addition to complying with the existing PHMSA regulatory requirements.

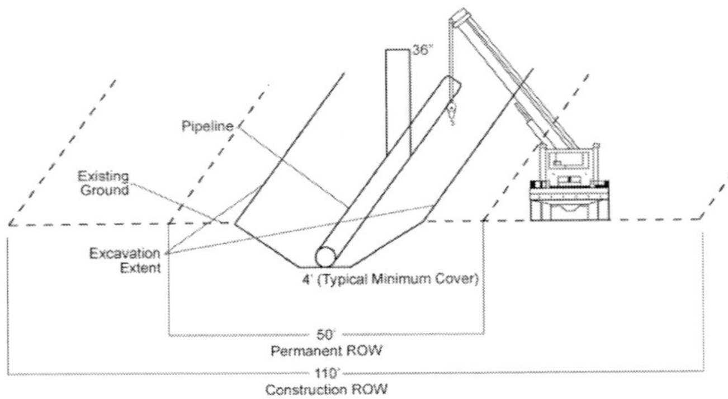

Figure ES-5. Pipeline Cross-section.

In consultation with PHMSA, DOS determined that incorporation of the Special Conditions would result in a Project that would have a degree of safety greater than any typically constructed domestic oil pipeline system under current regulations and a degree of safety along the entire length of the pipeline system that would be similar to that required in high consequence areas as defined in the regulations. Key aspects of the Special Conditions are summarized below. Appendix U of the EIS presents the Special Conditions and a comparison of the conditions with the existing regulatory requirements.

Pipe Design and Manufacturing

The first nine Special Conditions present design standards to be used in manufacturing the pipe and requirements for pipe materials, pipe inspections at the mill and in the field, performance tests, and quality control procedures.

System Design, Construction and Testing

Conditions 10 through 23 address design and construction of the proposed Project, including testing of Project components. Those Conditions present requirements for aspects of the proposed Project such as field coatings, depth of cover over the pipeline, temperature and overpressure control, welding procedures, and testing prior to operations. Testing requirements include hydrostatic testing, a process which involves filling the line with water and increasing the pressure within the pipeline to test the pipeline's ability to withstand pressure. If the test water pressure drops, further testing must be conducted and reported to PHMSA, and faulty pipeline sections must be repaired or replaced. Operations could not begin until the entire system has passed the required hydrostatic testing.

Operations, Maintenance, and Monitoring

Conditions 24 through 49 present the requirements for the Supervisory Control and Data Acquisition (SCADA) system that would be used to remotely monitor and control the pipeline, as well as requirements for internal corrosion inspection, cathodic protection, identification of the location of the pipeline with aboveground markers, internal pipeline inspections using electronic sensing devices termed "smart pigs," visual monitoring of the pipeline corridor, and repair procedures. The SCADA system would alert the Operations Control Center of an abnormal operating condition, indicating a possible release of oil. The system would include automatic features that would ensure operation within prescribed pressure limits. There would also be a complete backup system.

Pipeline pressure is the primary indicator used by the SCADA system to detect an oil spill. If the monitoring system identifies a pressure change in the pipeline, the controller would evaluate the data to determine if it is a false alarm or an actual spill. Using pipeline pressure allows the operator to detect leaks down to approximately 1.5 to 2 percent of pipeline flow rate.

The proposed Project would also include a computer-based system that does not rely on pipeline pressure to assist in identifying leaks below the 1.5 to 2 percent detection thresholds.

In addition to computer monitoring, there would be scheduled patrols of the pipeline right-of-way as well as public and landowner awareness programs. Communities along the pipeline would be given information to facilitate the reporting of suspected leaks and events that could suggest a threat to pipeline safety.

Reporting, Record Keeping, and Certification

The final eight conditions present requirements for maintaining detailed records, development a right-of-way management plan, reporting to PHMSA, and providing PHMSA with certification from a senior officer of Keystone that it has complied with the Special Conditions.

SPILL POTENTIAL AND RESPONSE

Spills could result from many causes, including corrosion (external or internal), excavation equipment, defects in materials or in construction, over-pressuring the pipeline, and geologic hazards, such as ground movement, washouts, and flooding. Although the leak detection system would be in place, some leaks might not be detected by the system. For example, a pinhole leak could be undetected for days or a few weeks if the release volume rate were small and in a remote area.

In most cases the oil from a small leak would likely remain within or near the pipeline trench where it could be contained and cleaned up after discovery. As a result, for most small leaks it is likely that the oil would be detected before a substantial volume of oil reaches the surface and affects the environment. Spills may be identified during regular pipeline aerial inspections, by ground patrols and maintenance staff, or by landowners or passersby in the vicinity of the spill.

For larger spills, the released oil would likely migrate from the release site. However, DOS analysis of previous large pipeline oil spills suggests that the depth and distance that the oil would migrate would likely be limited unless it reaches an active river, stream, a steeply sloped area, or another migration pathway such as a drainage ditch.

Figure ES-6. Smart Pig.

Estimated Frequency of Spills

In spite of the safety measures included in the design, construction, and operation of the proposed Project, spills are likely to occur during operation over the lifetime of the proposed Project. Crude oil could be released from the pipeline, pump stations, or valve stations.

Although a large spill could occur at the proposed Cushing tank farm, each of the three 350,000-barrel tanks would be surrounded by a secondary containment berm that would hold 110 percent of the contents of the tank plus freeboard for precipitation. Therefore, there would have to be a concurrent failure of the secondary containment berm for a tank-farm spill to reach the area outside of the tank. Such an event is considered unlikely.

DOS calculated estimates of spill frequency and spill volumes. Those estimates included potential spills from the pipeline, pump stations, and valve stations. The calculations used data from the PHMSA spill incident database for hazardous liquid pipelines and crude oil pipelines, and from the National Response Center (NRC) database for releases and spills of hazardous substances and oil.

Based on those data, DOS calculated that there could be from 1.18 to 1.83 spills greater than 2,100 gallons per year for the entire Project. The estimated frequency of spills of any size ranged from 1.78 to 2.51 spills per year.

Figure ES-7. Pump Station on the Existing Keystone Oil Pipeline System.

Keystone submitted a risk analysis that also included an estimate of the frequency of spills over the life of the proposed Project. Keystone's analysis was for the pipeline only and did not include releases from pump stations, valve stations, or the tank farm.

Keystone initially calculated a spill frequency of 1.38 spills per year based only on the historical PHMSA spill incident database available in 2008 when the application was submitted. Keystone also calculated a Project-specific spill frequency for the pipeline that considered the specific terrain and environmental conditions along the proposed Project corridor, required regulatory controls, depth of cover, strength of materials, and technological advances in the design of the proposed Project. Using those factors, Keystone estimated that there could be 0.22 spills per year from the pipeline.

Spills from the Existing Keystone Oil Pipeline System

The existing Keystone Oil Pipeline System has experienced 14 spills since it began operation in June 2010. The spills occurred at fittings and seals at pump or valve stations and did not involve the actual pipeline. Twelve of the spills remained entirely within the confines of the pump and valve stations. Of those spills, 7 were 10 gallons or less, 4 were 100 gallons or less, 2 were between 400 and 500 gallons, and 1 was 21,000 gallons.

The spill of 21,000 gallons occurred when a fitting failed at the Ludden, North Dakota pump station. As a result, PHMSA issued a Corrective Action Order, halting pipeline operation. Keystone was required to consult with PHMSA before returning the pipeline to operation. In that incident, most of the oil was contained within the pump station, but 210 gallons discharged from the pump station to adjacent land. The land affected was treated in place in compliance with North Dakota Department of Health land treatment guidelines.

Maximum Spill Volume

Keystone conducted an assessment of the maximum potential pipeline spill volume from a complete pipeline structural failure. Keystone estimated that the maximum spill volume would be approximately 2.8 million gallons, which would be possible along less than 1.7 miles of the proposed pipeline route due to topographic conditions. For approximately 50 percent of the proposed pipeline route (approximately 842 miles), the maximum spill volume would be approximately 672,000 gallons.

Emergency Planning

As required by PHMSA regulations, Keystone must submit an Emergency Response Plan and a Pipeline Spill Response Plan to PHMSA for review prior to initiation of operation of the proposed Project. These plans would not be completed until the final details of the proposed Project are established in all applicable permits.

If a leak is suspected, the Emergency Response Plan and Pipeline Spill Response Plan would be initiated. After confirmation that a spill occurred, the operator would shut down pumps and close the isolation valves, actions that would require approximately 12 minutes.

Local Emergency Planning Committees (LEPCs)

LEPCs were established as a part of the Emergency Planning and Community Right-to-Know Act. Keystone has committed to a communication program to reach out to LEPCs along the proposed pipeline corridor during development of the Emergency Response Plan and the Pipeline Spill Response Plan, with particular consideration given to emergency planning for low income and minority populations. The LEPCs would participate in emergency response consistent with their authority under the Right-to- Know Act and as required by their local emergency response plans.

Potential Environmental Impacts of Oil Spills

Impacts from an oil spill would be affected by variables such as the weather, time of year, water level, soil, local wildlife, and human activity. The extent of impact would also depend on the response time and capabilities of the emergency response team.

The greatest concern would be a spill in environmentally sensitive areas, such as wetlands, flowing streams and rivers, shallow groundwater areas, areas near water intakes for drinking water or for commercial/industrial uses, and areas with populations of sensitive wildlife or plant species.

General Types of Potential Impacts

There are two primary types of impacts that occur with a spill of crude oil – physical impacts and toxicological impacts. Physical impacts typically consist of the coating of soils, sediments, plants, and animals. The coating of organisms can result in effects such as preventing them from feeding or obtaining oxygen, reducing the insulating ability of fur or feathers, and adding weight to the organism so that it cannot move naturally or maintain balance. In addition, oil may coat beaches along rivers or lakes and foul other human-use resources.

Toxicological impacts of an oil spill are a function of the chemical composition of the oil, the solubility of each class of compounds in the oil, and the sensitivity of the area or organism exposed. Crude oil may be toxic when ingested. Ingestion typically occurs when an oiled animal attempts to clean its fur or feathers. Some of the possible toxic effects include direct mortality, interference with feeding or reproductive capacity, disorientation, reduced resistance to disease, tumors, reduction or loss of various sensory perceptions, and interference with metabolic, biochemical, and genetic processes.

Birds typically are the most affected wildlife due to an oil spill. Oil on feathers causes hypothermia or drowning due to the loss of flotation, and birds may suffer both acute and chronic toxicological effects. In addition, dead oiled birds may be scavenged by other animals.

Fish and aquatic invertebrates could also experience toxic impacts of spilled oil. The potential impacts would generally be greater in standing water habitats – such as wetlands, lakes and ponds – than in flowing rivers and creeks.

Crude oil spills are not likely to have toxic effects on the general public because of the many restrictions that local, state and federal agencies impose to avoid environmental exposure after a spill.

Potential Impacts to the Ogallala Aquifer and other Groundwater Areas

DOS recognizes the public's concern for the Northern High Plains Aquifer System, which includes the Ogallala aquifer formation and the Sand Hills aquifer unit.

The Northern High Plains Aquifer system supplies 78 percent of the public water supply and 83 percent of irrigation water in Nebraska and approximately 30 percent of water used in the U.S. for irrigation and agriculture. Of particular concern is the part of the aquifer which lies below the Sand Hills region. In that region, the aquifer is at or near the surface.

DOS assessed the potential impacts of the proposed Project on many aquifer systems. The aquifer analysis included the identification of potable groundwater in water wells within 1 mile of the proposed centerline of the pipeline. More than 200 Public Water Supply wells, most of which are in Texas, are within 1 mile of the proposed centerline, and 40 private water wells are within 100 feet of the centerline. No sole-source aquifers, or aquifers serving as the principal source of drinking water for an area, are crossed by the proposed pipeline route.

The potential for a crude oil spill to reach groundwater is related to the spill volume, the viscosity and density of the crude oil, the characteristics of the environment into which the crude oil is released (particularly the characteristics of the underlying soils), and the depth to groundwater. The depth to groundwater is less than 10 feet for about 65 miles of the proposed route in Nebraska and there are other areas of shallow groundwater in each state along the proposed route. Diluted bitumen and synthetic crude oil, the two types of crude oil that would be transported by the proposed Project, would both initially float on water if spilled. Over time, the lighter aromatic fractions of the crude oil would evaporate, and water-soluble components could enter the groundwater.

Studies of oil spills from underground storage tanks indicate that potential surface and groundwater impacts are typically limited to several hundred feet or less from a spill site. An example of a crude oil release from a pipeline system into an environment similar to the Northern High Plains Aquifer system occurred in 1979 near Bemidji, Minnesota.

While the conditions at Bemidji are not fully analogous to the Sand Hills region, extensive studies of the Bemidji spill suggest that impacts to shallow groundwater from a spill of a similar volume in the Sand Hills region would affect a limited area of the aquifer around the spill site. In no spill incident scenario would the entire Northern High Plains Aquifer system be adversely affected.

In addition to the Northern High Plains Aquifer system, there are other groundwater areas along the proposed route, including shallow or near-surface aquifers. DOS in consultation with PHMSA and EPA determined that Keystone should commission an independent consultant to review the Keystone risk assessment. The independent review will be conducted by a firm approved by DOS in concurrence with PHMSA and EPA, and would focus on a review of valve placement and the possibility of deploying external leak detection systems in areas of particularly sensitive environmental resources, but would not be limited to those issues. The specific scope of the analysis will be approved by DOS, PHMSA, and EPA. DOS, with concurrence from PHMSA and EPA, will determine the need for any additional mitigation measures resulting from the analysis.

Potential Environmental Justice Concerns

Low income and minority communities could be more vulnerable to health impacts than other communities in the event of a spill, particularly if access to health care is less available in the release area. Exposure pathways could include direct contact with the crude oil, inhalation of airborne contaminants, or consumption of food or water contaminated by either the crude oil or components of the crude oil. Keystone agreed to remediate spills, restore the affected areas, and provide alternative water supplies if a spill contaminates groundwater or surface water. Keystone also agreed to develop communications directed at bilingual communities, such as signage in both English and Spanish languages, and emergency communications in both languages.

Alternatives Considered

DOS considered the following three major alternative scenarios:

- No Action Alternative – potential scenarios that could occur if the proposed Project is not built and operated;
- System Alternatives – the use of other pipeline systems or other methods of providing Canadian crude oil to the Cushing tank farm and the Gulf Coast market;
- Major Route Alternatives – other potential pipeline routes for transporting heavy crude oil from the U.S./Canada border to Cushing, Oklahoma and the Gulf Coast market.

No Action Alternative

Under the No Action Alternative, the potential adverse and positive impacts associated with building and operating the proposed Project would not occur. However, there is an existing market demand for heavy crude oil in the Gulf Coast area. The demand for crude oil in the Gulf Coast area is projected to increase and refinery runs are projected to grow over the next 10 years, even under a low demand outlook.

A report commissioned by the Department of Energy (DOE) indicated that whether the proposed Project is built or not is unlikely to impact the demand for heavy crude oil by the Gulf Coast refineries. Even if improved fuel efficiency and broader adoption of alternative fuels reduced overall demand for oil, demand for Canadian heavy crude oil at Gulf Coast refineries would not be substantially affected.

At the same time, three of the four countries that are major crude oil suppliers to Gulf Coast refineries currently face declining or uncertain production horizons. As a result, those refineries are expected to obtain increased volumes of heavy crude oil from alternative sources in both the near term and further into the future. Implementation of the No Action Alternative would not meet this need.

If the proposed Project is not built and operated, Gulf Coast refineries could obtain Canadian crude oil transported through other new pipelines or by rail or truck transport. Other pipeline projects have been proposed to transport Canadian crude oil to the Gulf Coast area, and both rail transport and barge transport could be used to meet a portion of the need. In addition, the Gulf Coast refineries could obtain crude oil transported by marine tanker from areas outside of North America. Many of the sources outside of North America are in regions that are experiencing declining production or are not secure and reliable sources of crude oil, including the Middle East, Africa, Mexico, and South America.

As a result of these considerations, DOS does not regard the No Action Alternative to be preferable to the proposed Project.

If the proposed Project is not implemented, Canadian producers would seek alternative transportation systems to move oil to markets other than the U.S. Several projects have been proposed to transport crude oil out of using pipelines to Canadian ports.

Whether or not the proposed Project is implemented, Canadian producers would seek alternative transportation systems to move oil to markets other than the U.S. Several projects have been proposed to transport crude oil out of the oil sands area of Alberta using pipelines to Canadian ports.

System Alternatives

System alternatives would use combinations of existing or expanded pipeline systems, pipeline systems that have been proposed or announced, and non-pipeline systems such as tank trucks, railroad tank cars, and barges and marine tankers to transport Canadian heavy crude oil to Gulf Coast refineries.

None of the pipeline systems considered would be capable of transporting Canadian crude oil to Gulf Coast delivery points in the volumes required to meet Keystone's commitments for transporting 380,000 bpd to delivery points in Texas. Therefore they would not meet the purpose of the proposed Project. A combination of the pipeline systems considered could, over time, deliver volumes of Canadian oil sands crude oil in volumes similar to the volumes that would be transported by the proposed Project. However, that would not meet the near-term need for heavy crude oil at the Gulf Coast refineries. Expanding the pipeline systems that were considered to meet the purpose of the proposed Project or construction of new components or a combination of those systems would result in impacts similar to those of the proposed Project.

The trucking alternative would add substantial congestion to highways in all states along the route selected, particularly at and near the border crossing and in the vicinity of the delivery points. At those locations it is likely that there would be significant impacts to the existing transportation systems. Trucking would also result in substantially higher greenhouse gas emissions and a higher risk of accidents than transport by pipeline.

Development of a rail system to transport the volume of crude oil that would be transported by the proposed Project would likely produce less impact from construction than would the proposed Project because it could be done using existing tracks. However, there would be greater safety concerns and greater impacts during operation, including higher energy use and greenhouse emissions, greater noise impacts, and greater direct and indirect effects on many more communities than the proposed Project.

As a result of these considerations as described in Section 4.2 of the EIS, system alternatives were considered either not reasonable or not environmentally preferable.

Major Route Alternatives

The analysis of route alternatives considered 14 major route alternatives. Figure ES-8 depicts the alternative routes considered. The analysis of alternatives routes was conducted following the approach to assessments of alternative pipeline routes used by the Federal Energy Regulatory Commission. As a result, the analysis began with a screening process that first established criteria for screening alternatives, then identified potential alternatives that met the criteria, and determined whether or not they would (1) meet the purpose of and need for the proposed Project, and (2) be technically and economically practicable or feasible. For those alternatives meeting the criteria, DOS assessed whether or not the alternative offered an overall environmental advantage over the proposed route.

Due to public concern regarding the Ogallala Aquifer (Northern High Plains Aquifer system) and the Sand Hills region, 5 of the alternative routes were developed to either minimize the pipeline length over those areas or avoid the areas entirely. These alternative routes consisted of I-90 Corridor Alternatives A and B, Keystone Corridor Alternatives 1 and 2 (which are parallel to all or part of the route of the existing Keystone Oil Pipeline System), and the Western Alternative.

The assessment considered the environmental characteristics of the areas that these alternatives would cross, including the presence of aquifers, the depth of wells, developed land, forested areas, wetlands, and streams and rivers.

The Western Alternative was eliminated since it was financially impracticable. Although the other four route alternatives could have been eliminated based on consideration of economical and technical practicability and feasibility without further evaluation, they were nonetheless examined further with an emphasis on groundwater resources. The I-90 Corridor and Keystone Corridor alternatives would all avoid the Sand Hills; however, they would not avoid the Northern High Plains Aquifer system, and they would not avoid areas of shallow groundwater. Instead, these routes would shift risks to other areas of the Northern High Plains Aquifer system and to other aquifers.

In addition, these alternatives would be longer than the proposed route and would disturb more land and cross more water bodies than the proposed route. In addition, I-90 Corridor Alternatives A and B require crossing Lake Francis Case on the Missouri River which would pose technical challenges due to the width of the reservoir and the slope of the western side of the crossing area.

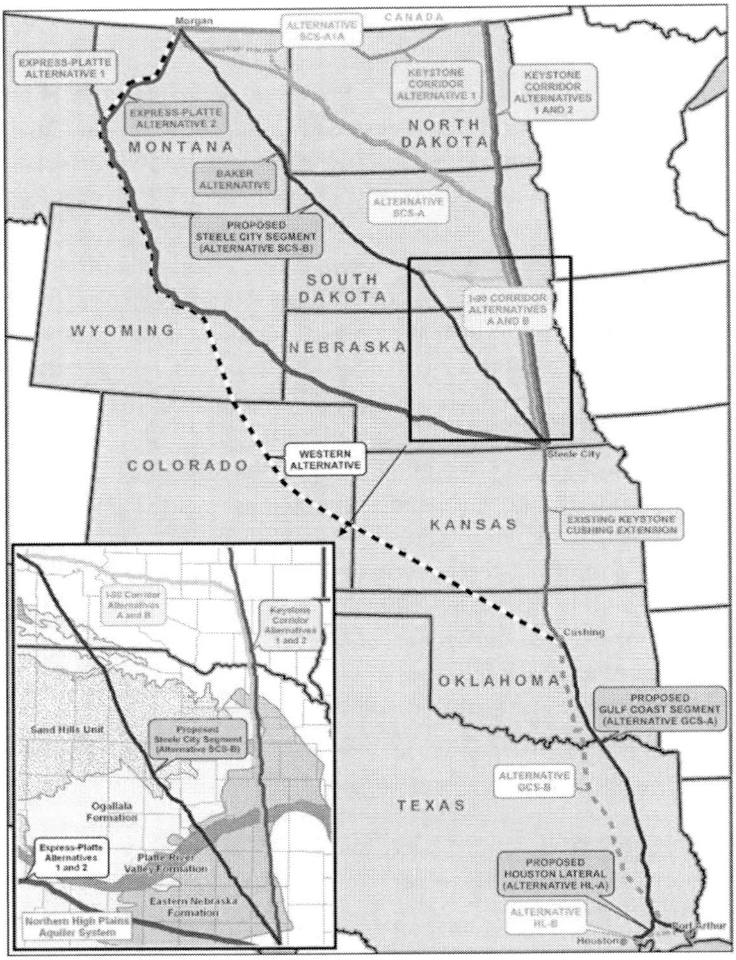

Figure ES-8. Major Route Alternatives.

Keystone Corridor Alternatives 1 and 2 would cost about 25 percent more than the proposed Project (about $1.7 billion more) and implementation of either of those alternatives would compromise the Bakken Marketlink Project and the opportunity to transport crude oil from the producers in the Bakken formation to markets in Cushing and the Gulf Coast.

Based on the above considerations and as described in Section 4.3 of the EIS, DOS eliminated the major potential route alternatives from further consideration.

Route Variations and Minor Realignments

A route variation is a relatively short deviation from a proposed route that replaces a segment of the proposed route. Variations are developed to resolve landowner concerns and impacts to cultural resource sites, wetlands, recreational lands, and terrain.

DOS consulted with the Bureau of Land Management and state agencies to negotiate route variations and minor realignments, including nearly 100 in Montana and about 240 minor realignments in other states along the proposed route. Additional route variations and minor realignments may be added in response to specific conditions that may arise throughout the construction process.

The variations and minor realignments would replace short segments of the proposed Project, are relatively close to the proposed route, and would be implemented in accordance with applicable regulatory requirements of federal, state, or local permitting agencies. DOS considers the variations and minor realignments selected to have been evaluated sufficiently to meet the environmental review requirements of the National Environmental Protection Act.

Other Alternatives Considered

DOS also considered several other scenarios in response to comments on the draft EIS. The alternative pipeline designs considered consisted of an aboveground pipeline and a smaller diameter pipe to decrease the volume of oil released from a spill. DOS also considered alternative sites for the major aboveground facilities of the proposed Project, including pump stations, mainline valves, and the Cushing tank farm. None of the alternative designs or facility locations were considered safer or environmentally preferable to the proposed Project design.

Agency Preferred Alternative

DOS did not find any of the major alternatives to be preferable to the proposed Project for the reasons presented in the final EIS and summarized above. As a result, the agency-preferred alternative is the proposed Project route with the variations and minor route realignments described in the EIS, and the proposed location of the Cushing tank farm.

Environmental Analyses

Four levels of impact duration were considered in the analysis of potential environmental impacts due to construction and normal operation of the proposed Project: temporary, short-term, long-term, and permanent. Temporary impacts generally occur during construction, with the resources returning to pre-construction conditions almost immediately afterward. Short-term impacts could continue for approximately 3 years after construction, and impacts were considered long term if the resources would require more than 3 years to recover. Permanent impacts would occur if the resources would not return to pre-construction conditions during the life of the proposed Project, such as impacts to land use due to installation of pump stations.

Conclusions in the EIS are based on the analysis of environmental impacts and the understanding that:

- Keystone would comply with all applicable laws and regulations;
- The proposed Project would be constructed, operated, and maintained as described in the EIS;
- Keystone has agreed to incorporate the 57 Project-specific Special Conditions developed by PHMSA into the proposed Project;
- Keystone has agreed to implement the measures designed to avoid or reduce impacts described in its application for a Presidential Permit and supplemental filings with DOS, the measures in its Construction, Mitigation, and Reclamation (CMR) Plan presented in Appendix B of the EIS, and the construction methods for the Sand Hills region described in Appendix H to the EIS; and
- Keystone would incorporate the mitigation measures required in permits issued by environmental permitting agencies into the construction, operation, and maintenance of the proposed Project.

Environmental Justice

Executive Order 12898 requires federal agencies to address and mitigate potential adverse impacts to minority and low income populations. In consultation with EPA, DOS identified these communities within a 4-mile-wide corridor centered on the pipeline using census and county level data.

Potential Construction Impacts

The assessment suggested that potential impacts to minority and low income populations could occur primarily in Harris, Jefferson, and Angelina Counties in Texas and in Lincoln County, Oklahoma. During construction, potential impacts include exposure to increased dust and noise, disruption of traffic patterns, and increased competition for social services in underserved populations. At any given location along the proposed pipeline route, the duration of the construction period would typically range from 20 to 30 working days. As a result, the impacts to minority and low-income populations due to construction would be temporary and minor.

Medical Services

Areas along the pipeline route that are medically underserved may be more vulnerable during construction periods. These communities have been identified as Health Professional Shortage Areas or Medically Underserved Areas/Populations. However, construction-related disruptions in those areas would be temporary and minor. In areas in Montana and South Dakota, minor medical needs of workers would be handled in construction camps to avoid or minimize the need for medical services from the surrounding communities.

Air Emissions Related to Environmental Justice Issues

The refineries that are likely to receive oil transported by the pipeline are already configured to process heavy crude oil, and in the future would seek to continue processing heavy crude oil whether or not the proposed pipeline is constructed. The analysis in the EIS, including a DOE-commissioned study, indicates that the proposed Project would not likely affect the overall quality or quantity of crude oil refined in the Gulf Coast region, and, as a result, would not likely effect refinery emissions.

Greenhouse Gas Emissions

DOS commissioned a detailed study of greenhouse gas life-cycle emissions that compared Canadian oil sands crude with other selected reference crudes. This study was a thorough review of recent scientific literature on greenhouse gas life-cycle emissions for Canadian oil sands crude including extraction, upgrading, transportation, refining, and combustion.

The study's major conclusion was that, throughout its life cycle, oil sands crude is, on average, more greenhouse gas intensive than the crude oil it would replace in the U.S. However, the relative greenhouse gas intensity varies depending on (1) study design factors, such as the reference crudes selected for comparison with Canadian oil sands crudes (e.g., 2005 U.S. average crude oil, Venezuelan Bachaquero, Middle East Sour, and Mexican Heavy) and the timeframe selected, and (2) study assumptions, such as the extraction method and the mix of crudes that would be transported by the pipeline.

For example, the Department of Energy's National Environmental Technology Lab (NETL) study indicated that the life-cycle greenhouse gas emissions of gasoline produced from Canadian oil sands crude are approximately 17 percent higher than gasoline from the 2005 average mix of crude oil consumed in the U.S. The NETL study serves as a key input for analyses conducted by EPA and DOE. In comparison, a study conducted by TIAX, LLC, found that the greenhouse gas emissions from gasoline produced from Canadian oil sands crude are only 2 percent higher when compared to gasoline from Venezuelan heavy crude, a type of crude oil that is similar to the crude oil that would be transported by the proposed Project and is currently refined in large quantities by Gulf Coast refineries.

The proposed Project is not likely to impact the amount of crude oil produced from the oil sands. However, for illustrative purposes, the DOS commissioned study estimated that incremental lifecycle U.S. greenhouse gas emissions from displacing reference crude oils with Canadian oil sands crude oils imported through the proposed Project would be between 3 and 21 million metric tons of carbon dioxide emissions annually. This range is equivalent to annual greenhouse gas emissions from the combustion of fuels in 588,000 to 4,061,000 passenger vehicles.

In addition, current projections suggest that the amount of energy required to extract all crude oils is projected to increase over time due to the need to extract oil from ever deeper reservoirs using more energy intensive techniques. However, while the greenhouse gas intensity of reference crude oils may trend upward, the projections for the greenhouse gas intensity of Canadian oil sands crude oils suggests that they may stay relatively constant. Although there is some uncertainty in the trends for both reference crude oils and oil sands derived crude oils, on balance it appears that the gap in greenhouse gas intensity may decrease over time.

Geology and Soils

Geologic Hazards

Potential geologic hazards assessed in the EIS include seismic hazards (earthquakes), landslides, or subsidence (sink holes). The proposed route extends through relatively flat and stable areas and the potential for these events is low. The pipeline would not cross any known active faults with confirmed surface offsets. During construction, land clearing could increase the risk of landslides and erosion. Keystone agreed to construct temporary erosion control systems and revegetate the right-of-way after construction.

There is a risk of subsidence (sink holes) where the proposed route potentially crosses karst formations in Nebraska, Oklahoma, and Texas. Site-specific studies would be conducted as necessary to characterize the karst features, if they are encountered, and evaluate and modify construction techniques as necessary in these areas. The overall risk to the pipeline from karst-related subsidence is expected to be minimal.

Soils and Sediments

Potential impacts to soils include soil erosion, loss of topsoil, soil compaction, soil contamination, damage to existing tile drainage systems, and permanent increases in the proportion of large rocks in the topsoil. However, Keystone agreed to construction procedures that are designed to reduce the likelihood and severity of Project impacts to soils and sediments, including topsoil segregation methods, and to mitigate impacts to the extent practicable.

Sand Hills Region

Of particular concern is the soil of the Sand Hills region of Nebraska, which is particularly vulnerable to wind erosion. To address this concern, Keystone developed and agreed to construction, reclamation, and post-construction procedures specifically for this area in consultation with local experts and state agencies. The goal of the Sand Hills region reclamation plan is to protect this sensitive area by maintaining soil structure and stability, stabilizing slopes to prevent erosion, restoring native grass species, and maintaining wildlife habitat and livestock grazing areas. Keystone agreed to monitor the right-of-way through the Sand Hills region for several years to ensure that reclamation and revegetation efforts are successful.

Water Resources

Groundwater

Many of the aquifers along the proposed route are isolated from the surface due to soil types above the aquifers that prevent or slow downward migration of water. However, shallow or near-surface aquifers are also present along the proposed route, as discussed above. Construction of the proposed Project may result in temporary to shortterm increases in suspended solids in the shallow aquifers. The risk of dewatering shallow groundwater aquifers during construction or reducing groundwater quality due to increased sediments in the water would be temporary to short term.

At some locations, groundwater may be used as a source of water for pressure testing the pipeline during construction. Keystone must obtain all applicable water withdrawal and discharge permits prior to testing, and the test water would be tested and discharged in accordance with permit requirements.

River and Stream Crossings

Surface water bodies would be crossed using one of three methods: the open-cut wet method, the dry-cut method, or the horizontal directional drilling method. The method selected would be based on the characteristics of the crossing location and the requirements of the permitting agencies.

The open-cut wet method, which involves trenching while the stream is flowing, would result in temporary increases in turbidity and bank erosion where vegetation is removed. The dry-cut method, which involves diverting stream flow around the construction site, results in lower increases in turbidity than the open-cut wet method.

Horizontal directional drilling would minimize impacts to the stream or river because it involves drilling well below the streambed. This method would be selected at large body crossings to avoid disturbing the streambeds and streamflow and to reduce the potential that deep scour during flooding would endanger pipeline integrity. Figure ES-9 presents a cross section of a river crossing using the horizontal directional drilling method.

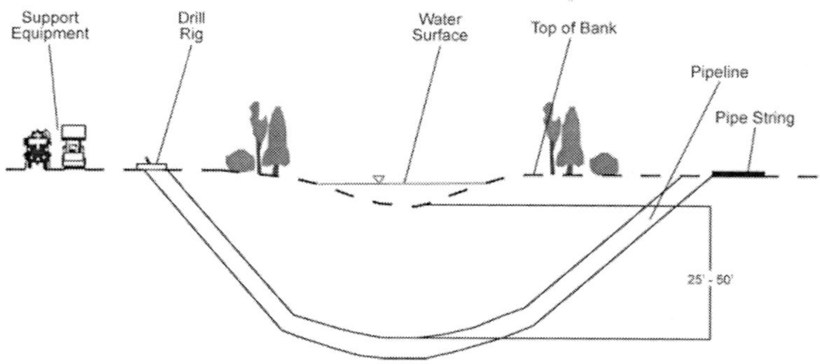

Figure ES-9. Cross Section of Horizontal Directional Drilling Method.

At all water crossings, Keystone agreed to use vegetative buffer strips, drainage diversion structures, and sediment barriers, and limit vegetation clearing to reduce siltation and erosion. After construction, the right-of-way would be restored and revegetated to reduce the potential for erosion of the stream bank.

Hydrostatic Test Water

Water used to pressure test the pipeline during construction would be discharged to its source waters or to an approved upland area within the same drainage and tested to ensure it meets applicable water quality standards and discharge rates.

Wetlands

The proposed Project route crosses emergent, scrub/shrub, and forested wetlands that are protected by the U.S. Army Corps of Engineers (USACE) and applicable state agencies under the review of EPA through Section 401 and 404 of the Clean Water Act. Specific plans regarding wetland avoidance and minimization of impacts, and the development of mitigation to compensate for the permanent loss or conversion of forested to emergent wetlands would be further developed during the permitting process. Wetland impacts presented in the EIS represent preliminary estimates based on the best available wetland information. DOS reviewed potential impacts to wetlands and the avoidance, minimization, and mitigation process that would be followed with USACE and EPA.

Most wetlands crossed by the proposed Project in Montana, South Dakota, and Nebraska are emergent wetlands, and most wetlands crossed by the proposed Project in Oklahoma and Texas are forested wetlands. Construction of the pipeline would affect wetlands and their functions primarily during and immediately after construction activities, but permanent changes also are possible. Keystone agreed to use construction methods that avoid or minimize impacts to wetlands. These measures include installing trench breakers and/or sealing the trench to maintain the original wetland hydrology to avoid draining wetlands, using timber mats to protect wetlands during construction, and restoring wetland areas to a level consistent with the requirements of the applicable permits.

Most wetland vegetation communities would transition back into a community that would function similarly to the previously undisturbed wetland. Because most wetlands would be restored, the overall impact of the proposed Project to wetlands would be minor to moderate and would range in duration from short term to the life of the proposed Project. However, some forested and scrub-shrub wetlands over the pipeline would be converted to herbaceous wetlands since trees and shrubs would not be allowed to grow over the pipeline for inspection and integrity purposes. Keystone is working with each USACE district along the proposed route to identify wetlands and to develop wetland mitigation and compensation plans for the permanent conversion of forested wetland to herbaceous wetland.

Texas Bottomland Hardwood Wetlands

These are forested wetlands with trees, such as Bald Cypress, Water Oak, Water Hickory, and Swamp Tupelo that can exist in lowland floodplains in the Gulf Coast states. Clearing bottomland hardwood trees during construction would result in long-term to permanent impacts because forests require decades to reestablish and would mature over the span of centuries. DOS reviewed potential Project impacts on bottomland hardwood wetlands with EPA and USACE. Preliminary mitigation measures to protect bottomland hardwood wetlands are discussed in the EIS and would be developed further by the USACE during the wetland permitting process.

Terrestrial Vegetation

The proposed Project crosses primarily grasslands and rangelands, followed by croplands, upland forests, developed lands, and wetlands. After construction, Keystone agreed to restore topsoil, slopes, contours, and drainage patterns to preconstruction conditions as practicable and to reseed disturbed areas to restore vegetation cover, prevent erosion, and control noxious weeds. Keystone committed to controlling the introduction and spread of noxious weeds and pests by adhering to construction and restoration procedures recommended by local, state, and federal agencies. Soils and vegetation over the pipeline would be warmed slightly compared to surrounding soils by heat loss from the pipeline during operation.

Native Grasslands and Rangelands

Native mixed shrub rangelands would be crossed by the proposed Project in Montana and South Dakota and native grasslands would be crossed by the proposed Project in the Sand Hills region in Nebraska. Both of these native prairie habitats would be challenging to reclaim. In recognition of these challenges, Keystone developed specific construction and reclamation methods for the proposed Project in consultation with local, state, and federal agencies and local experts to ensure that sagebrush and native grasses are restored to rangelands in Montana and South Dakota and that fragile soils and diverse native vegetation cover are re-established in the Sand Hills region of Nebraska.

Upland and Riparian Forests

Native forests, especially forested floodplains, were once an integral component of the landscape throughout the Great Plains and they provide important habitats for wildlife. Clearing trees in upland and riparian forest communities would result in long-term impacts because trees would be required to remain outside of the 50-foot-wide permanent right-of-way. These impacts would last throughout the life of the proposed Project because trees would not be allowed to reestablish within the permanent right-of-way and because forests require decades to re-establish and would mature over the span of centuries.

Wildlife

Big game animals, small game animals and furbearers, waterfowl and game birds, and other nongame animals use habitats in and around the six states crossed by the proposed Project. Construction would result in the temporary and permanent loss and alteration of habitats which provide foraging, cover, and breeding habitats for wildlife. Most habitat loss would be temporary as vegetation cover would be reestablished after construction and would be small in context to habitats available throughout the region crossed by the proposed Project. Loss of shrublands and wooded habitats would be long-term (from 5 to 20 years or more), however; and trees and tall shrubs would not be allowed to re-establish over the pipeline for inspection and integrity purposes. Aboveground facilities would result in some permanent habitat loss. Power lines to pump stations can provide vantage perches for raptors that lead to increased predation on ground nesting birds and small mammals. Construction can produce short-term barriers to wildlife movement, direct and indirect mortality, and reduced survival and reproduction. Disturbance from construction activities may have moderate local affects on wildlife if important remnant habitats are crossed or when sensitive breeding or overwintering periods are not avoided. Habitat alteration and fragmentation caused by the pipeline right-of-way may reduce habitat suitability and use by wildlife.

Construction could also produce short-term barriers to wildlife movement, direct and indirect mortality, and reduced survival and reproduction. Disturbance from construction activities would have moderate local affects on wildlife if important remnant habitats are crossed or when sensitive breeding or overwintering periods are not avoided. Habitat alteration and fragmentation caused by construction of the pipeline could reduce habitat suitability and use by wildlife.

Figure ES-10. Texas Bottomland Hardwood Wetland.

Figure ES-11. Sand Hills Grassland.

During the environmental review of the proposed Project, state and federal wildlife management agencies were contacted and they provided information on sensitive seasons and wildlife habitats such as big game overwintering habitats, important riparian corridors, and raptor and other migratory bird nesting habitats. In addition state and federal wildlife management agencies provided recommendations for surveys to more specifically locate areas such as raptor nests and prairie dog colonies that could potentially be avoided. Keystone is working with state and federal wildlife management agencies to minimize impacts to wildlife during sensitive breeding periods. Measures developed to minimize impacts to wildlife include development of a Migratory Bird Conservation Plan in consultation with the USFWS, removal of litter and garbage that could attract wildlife, control of unauthorized off-road vehicle access to the construction right-of-way, and reclamation of native range with native seed mixes. Overall, the impact of construction to wildlife is expected to be minor and would be primarily temporary to short term. Normal Project operation would result in negligible effects to wildlife.

Keystone must work with state and federal wildlife management agencies to minimize impacts to wildlife during sensitive breeding periods. Overall, the impact of construction to wildlife is expected to be minor and would be primarily temporary to short term. Normal Project operation would result in negligible effects to wildlife.

Figure ES-12. Mule Deer.

Fisheries Resources

The proposed route would cross rivers and streams, including perennial streams that support recreational or commercial fisheries. Most potential impacts to fisheries resources would occur during construction and would be temporary to short term. Potential impacts from construction of stream crossings include siltation, sedimentation, bank erosion, sediment deposition, short-term delays in movements of fish, and transport and spread of aquatic invasive animals and plants. Keystone has agreed to minimize vehicle contact with surface waters and to clean equipment to prevent transportation of aquatic invasive animals and plants on equipment.

Most streams would be crossed using one of several trenching methods. Trenching stream crossings when water is still flowing through the stream bed can result in destruction of fish that do not avoid the construction area. Trenching methods may also use dams, pumps, and flumes to divert the stream flow around the trench location to allow a "dry" trenching method. However, direct disturbance to the stream bed can release fine sediments during construction through flowing waters or after the flow is returned to the stream bed. Sediment would be transported downstream and could affect fish, other aquatic life, and aquatic habitats through either direct exposure or smothering. Most stream crossings would be completed in less than 2 days, grading and disturbance to waterbody banks would be minimized, and crossings would be timed to avoid sensitive spawning periods, such that resulting steam bed disturbance and sediment impacts would be temporary and minor.

Most large rivers would be crossed using the horizontal directional drilling method which would install the pipeline well below the active river bed. As a result, direct disturbance to the river bed, fish, aquatic animals and plants, and river banks would be avoided. Keystone has developed site specific plans for horizontal directional drill crossings and has agreed to develop site-specific contingency plans to address unintended releases of drilling fluids that include preventative measures and a spill response plan.

Threatened and Endangered Species

The U.S. Fish and Wildlife Service (USFWS) is responsible for protecting threatened and endangered species under the Endangered Species Act (ESA). Federally-protected threatened or endangered species that are known or thought to be in the vicinity of the proposed Project include three mammals, five birds, one amphibian, five reptiles, three fish, two invertebrates, and four plants. DOS prepared a Biological Assessment and consulted with USFWS to evaluate the proposed Project's potential impact on federally-protected threatened or endangered species.

USFWS has determined that the proposed Project would have no affect on 12 of the listed species, and may affect, but is not likely to adversely affect 10 of those species. These evaluations are based on species occurrence and conservation measures developed in consultation with USFWS that Keystone has agreed to implement. DOS and USFWS determined that the proposed Project would likely adversely affect the American burying beetle and a formal consultation was initiated to determine whether impacts could jeopardize the continued existence of the species and to further develop conservation measures and an incidental take statement. Based on the formal consultation, USFWS is formulating a Biological Opinion that would be required prior to the issuance of a Record of Decision by DOS or any other federal cooperating agency.

Direct impacts to beetles could occur due to habitat loss, construction, and pre-construction conservation measures (where beetles would be trapped and relocated away from the project area). During operation, the flow of oil through the pipeline would generate heat that would warm the surrounding soils and could affect beetles during the winter when they bury themselves in the soil to hibernate. During formal consultation with the USFWS, conservation measures were developed that include Keystone providing funding for conservation efforts and monitoring of American burying beetle habitat restoration, and the establishment of a performance bond for supplemental habitat reclamation if initial reclamation efforts are unsuccessful.

Several candidate species for federal protection under the ESA are known or thought to be in the vicinity of the proposed Project including three birds, one reptile, one fish, and two plants. Measures that have been developed to avoid and minimize potential impacts to these species include reclamation of native range with native seed mixes, development of a Migratory Bird Conservation Plan in consultation with USFWS, and development of greater sage-grouse mitigation implementation plans for Montana and South Dakota in consultation with state and federal agencies.

A total of 35 state-protected species may also be present along the proposed right-of-way. These species have been designated by state wildlife management agencies as being of concern to assist with conservation planning and maintenance of the state's natural heritage. Conservation measures developed in consultation with state agencies include conducting additional species-specific surveys to determine whether nests, dens, or suitable habitats are present along the proposed right-of-way; adhering to construction timing restrictions to avoid the breeding, denning, and spawning seasons; and reducing the width of the construction right-of-way in areas where state-protected plant populations have been identified.

Cultural Resources

DOS, in coordination with consulting parties, has minimized the potential for adverse effects to historic properties along the Area of Potential Effect (APE) of the proposed Project by the development of avoidance and mitigation measures. Since 2008, DOS has consulted with Indian tribes, State Historic Preservation Officers, federal agencies and local agencies under Section 106 of the National Historic Preservation Act. As part of this effort, DOS initially contacted over 95 Indian tribes to find out their level of interest in becoming a consulting party. DOS also conducted Section 106 government-to-government consultation with the consulting parties for the proposed Project. DOS also invited the consulting tribes to prepare Traditional Cultural Property studies as part of the lead agency responsibilities for the identification, evaluation and mitigation of historic properties.

Figure ES-13. Recreational Fishing.

Figure ES-14. American Burying Beetle.

A Programmatic Agreement was developed by DOS and the parties. The Programmatic Agreement establishes a procedure for the further identification, evaluation, mitigation, and treatment of historic properties and will be completed prior to construction of the proposed Project. The Advisory Council on Historic Preservation participated in the development of this agreement with DOS and the other consulting parties. As part of this agreement, a Tribal Monitoring Plan and a Historic Trails and Archaeological Monitoring Plan were also developed. If previously unidentified archaeological sites are encountered during construction of the proposed Project, Keystone, DOS, and the consulting parties would follow the procedures described in the Unanticipated Discovery Plans.

Air Quality and Noise

Air Quality

Air quality impacts from construction would include emissions from construction equipment, temporary fuel transfer systems, fuel storage tanks, and dust and smoke from open burning. Most of these emissions would occur only intermittently, would be limited to active construction areas, and would be controlled to the extent required by state and local agencies.

All pump stations will be electrically powered by local utility providers. As a result, during normal operation there would be minor emissions from valves and pumping equipment at the pump stations. There would also be low levels of emissions from mobile sources, and low levels of emissions from the proposed Cushing tank farm and the surge relief systems at the delivery points. The proposed Project would not cause or contribute to a violation of any federal, state, or local air quality standards and it would not require a Clean Air Act Title V operating permit.

The proposed Project would cross five counties where the background concentration of ozone is greater than the national ambient air quality standards. Those areas are designated as nonattainment for the federal 8-hour ozone standard. However, the emissions from the proposed Project would be consistent with state implementation plans for air quality issues.

Noise

During construction there would be intermittent, temporary, and localized increases in sound levels as construction activities move through an area. To reduce construction noise impacts, Keystone agreed to limit the hours during which activities with high-decibel noise levels are conducted in residential areas, require noise mitigation procedures, monitor sound levels, and develop site-specific mitigation plans to comply with regulations. As a result, the potential noise impacts associated with construction would be minor and temporary.

During operation, sound levels within 2,300 feet of pump stations would increase. Outside of this distance, noise levels would remain at existing sound levels. Keystone committed to performing a noise assessment survey and to mitigating identified impacts by installing noise reducing measures at the pump stations.

Land Use, Recreation, and Visual Resources

The majority of land that would be affected by the project is privately owned (21,333 acres) with nearly equal amounts of state (582 acres) and federal (579 acres) lands being impacted.

Agriculture

After construction, nearly all agricultural land and rangeland along the right-of-way would be allowed to return to production with little impact on production levels in the long term. However, there would be restrictions on growing woody vegetation and installing structures within the 50-foot-wide permanent right-of-way. Keystone has agreed to compensate landowners for crop losses on a case-by-case basis.

There are 102 tracts of land that would be impacted which are part of the Conservation Reserve Program. The proposed Project is not expected to affect landowner ability to participate in that program.

Keystone agreed to use construction measures designed to reduce impacts to existing land uses, such as topsoil protection, avoiding interference with irrigation systems except when necessary, reducing construction time in irrigated areas, repairing or restoring drain tiles, restoring disturbed areas with custom seed mixes to match the native plants, providing access to rangeland during construction, installing temporary fences with gates around construction areas to prevent injury to livestock or workers, providing trench crossing areas to allow livestock and wildlife to cross the trench safely, and controlling noise and dust control.

Recreation

Operation of the proposed Project would not affect recreational resources, national or state parks, or users of those resources. Keystone has committed to cooperating with private landowners, and with federal, state, and local agencies to reduce the conflict between recreational users and Project construction.

Visual Resources

During construction, there would be visual impacts associated activities along the proposed right-of-way such as clearing, trenching, pipe storage, and installing above-ground structures. Most of the visual impacts of the pipeline corridor in agricultural and rangeland areas would be substantially reduced with restoration and revegetation. Keystone agreed to install vegetative buffers around the pump stations to reduce the visual impacts of those facilities. Overall, the visual impacts of the proposed Project would generally be minor to moderate.

Socioeconomics

During construction, there would be temporary, positive socioeconomic impacts as a result of local employment, taxes on worker income, spending by construction workers, and spending on construction goods and services. The construction work force would consist of approximately 5,000 to 6,000 workers, including Keystone employees, contractor employees, and construction and environmental inspection staff. That would generate from $349 million to $419 in total wages. An estimated $6.58 to $6.65 billion would be spent on materials and supplies, easements, engineering, permitting, and other costs.

Adverse impacts during construction could include temporary and minor increases in the need for public services, disruption of local transportation corridors, and reduced availability of transient housing. Keystone would establish four temporary work camps in southeastern Montana and northwestern South Dakota to minimize impacts to transient housing and public services in those areas. Operation of the proposed Project would also result in long-term to permanent beneficial socioeconomic impacts, including employment and income benefits resulting from long-term hires and local operating expenditures, and increased property tax revenues. An estimated $140.5 million in annual property tax revenues would be generated by the proposed Project.

Cumulative Impacts

The analysis of cumulative impacts combined the potential impacts of the proposed Project with the impacts of past, present, and reasonably foreseeable future actions in the vicinity of the proposed route. This assessment included consideration of the many existing pipelines, electrical transmission lines, and roadways, as well as other linear projects that are under construction, planned, proposed, or reasonably foreseeable in the vicinity of the proposed route. The analysis also included existing and likely energy development projects.

During construction, the proposed Project would contribute to cumulative dust and noise generation, loss of vegetation or crop cover, and minor localized traffic disruptions where other linear projects are under construction at the same time and are in the vicinity of the proposed route.

One of the primary contributions to cumulative effects during operation would be emissions from storage tanks. However, the proposed Project and all other petroleum storage projects would have to comply with the emissions limitations of air quality permits. In addition, where Project-related aboveground facilities and visible corridors are present along with those of other projects, there would be cumulative effects to visual resources. Other cumulative impacts associated with operation include changes in land use, terrestrial vegetation, wetland function, and wildlife habitat, as well as increases in tax revenues, and employment. Where the pump stations or compressor stations of other pipeline systems are in the vicinity of the pump stations for the proposed Project, there would also be cumulative noise impacts.

An increase in the development of wind power projects in the central plains region as well as increased need for electrical power is likely to increase the number of electrical transmission lines in the vicinity of the proposed route. If the construction of power distribution or transmission lines in the vicinity of the proposed route overlaps with construction of the proposed Project, short-term cumulative impacts associated with noise, dust, and general construction activity could occur. Likely cumulative impacts of the proposed Project and operation of new transmission lines include viewshed degradation, changes to land uses and vegetation, and impacts to birds.

Environmental Impacts in Canada

An evaluation of the impacts resulting from extraction of crude oil from the oil sands in Canada is outside of the scope of analysis required under the National Environmental Policy Act. However, in response to comments and as a DOS policy decision, the general regulatory oversight and the environmental impacts in Canada related to oil sands production were summarized in the EIS.

The potential environmental effects of the proposed Project have been assessed on both sides of the international border. In March 2010, the National Energy Board of Canada determined that the proposed Keystone XL Project is needed to meet the present and future public convenience and necessity, provided that the Board's terms and conditions presented in the project certificate are met. The Board's assessment included evaluations of need, economic feasibility, potential commercial impacts, potential environmental and socioeconomic effects, appropriateness of the general route of the pipeline, potential impacts on Aboriginal interests, and other issues.

Oil sands development projects undergo an environmental review in Canada under Alberta's Environmental Protection and Enhancement Act and other environmental regulations. Other federal and provincial agencies may participate in the review as Responsible Authorities or as Federal Authorities with specialist advice. Government regulators of oil sands activities in Canada are working to manage and provide regional standards for air quality, land impact, and water quality and consumption based on a cumulative effects approach.

Oil sands mining projects have reduced greenhouse gas emissions intensity by an average of 39 percent between 1990 and 2008 and are working toward further reductions. In addition, the Alberta Land Stewardship Act supports the Land-use Framework, which includes province-wide strategies for establishing monitoring systems, promoting efficient use of lands, reducing impact of human activities, and including aboriginal people in land-use planning.

In: The Keystone XL Pipeline Project ISBN: 978-1- 62081-221-1
Editors: G. B. Flynn and D. Burnett

Chapter 4

THE KEYSTONE XL PIPELINE*

Kerri-Ann Jones

Moderator: Good afternoon. We're pleased today to have Assistant Secretary Dr. Kerri-Ann Jones in the Bureau of Oceans and International Environmental and Scientific Affairs, who will discuss the Keystone XL Pipeline project. A reminder that today's call is on-the-record. Dr. Jones will have some brief opening remarks, followed by any questions that you might have. Without further ado, I'll turn it over to Dr. Jones for her opening remarks.

Assistant Secretary Jones: Good afternoon, everyone. Thank you for being on the call. We wanted to talk today about the announcement that we have made. As you know, today the Department of State recommended to President Obama that the presidential permit for the proposed Keystone XL pipeline be denied and that at this time the TransCanada Keystone XL pipeline be determined to not serve the national interest. The President agreed with the Department's recommendation. This recommendation was predicated on the fact that the Temporary Payroll Tax Cut Continuation Act that was passed in December does not provide sufficient time to obtain the information that we think is necessary to assess whether the project, in its current state, is in the national interest.

---

* This is an edited, reformatted and augmented version of the Briefing Transcript as issued by the U.S. Department of State, dated January 18, 2012.

As you – many of you know, we made a decision on November 10th that we needed additional information regarding this project, specifically about alternative routes through Nebraska. And we base this decision now on the fact that we don't have time to get that information, information we think is essential for making a well-informed decision. So I think I'll stop there and open it up for questions.

Moderator: Okay, Melissa. We're ready for questions.

Operator: Yes, sir. Thank you. As a reminder, to ask a question, please press *1 at this time. The first question we have comes from Josh Rogin, Foreign Policy. Your line's open.

Question: Yes, thank you very much for taking the time to do the call. I'm wondering, since you're saying that, the – there wasn't enough time to properly evaluate the proposal, was that evaluation made when the law was passed, or was – what have you been doing for the last – since the law was passed? Have you been working on this? Do you anticipate evaluating a new proposal? And could you also please characterize the level of White House involvement in the State Department's process during this period? Thank you.

Assistant Secretary Jones: Okay, that's a lot of questions.

Question: Yes, it is.

Assistant Secretary Jones: Let me start with the first one: What have we been doing since this Act was passed? I think you have been following our press announcements and our different comments. We have been analyzing the legislation. But when it was first proposed, we did mention and we did sort of confirm that we felt that any sort of imposition of a deadline would compromise the process. Our decision that we needed the extra information that we identified on November 10th wasn't changed by the legislation. And the legislation provided 60 days to really get all that information and do the analysis, and we just didn't think that was adequate and we couldn't really make a decision. And therefore, the pipeline would not – as it is currently presented, could not be deemed in the national interest. So that's number one.

I think the second one was more about the White House involvement, and there's sort of two phases here. The first phase, certainly the White House was not involved. November 10th we made this decision that we needed additional information and we informed the White House of that, and I think we talked about that in one of my previous press calls. Since the legislation was passed, we have consulted with the White House as it really talks about what the President needs to do, and we have been consulting with the White House because of the new legislation.

Question: So the point is that the charge against the State Department is that this was a White House-driven political decision. And so what I'm trying to get at here is, did this decision come from the White House or was this decision actually made inside the State Department?

Assistant Secretary Jones: This is a process decision that we made based on the decision that we made on November 10th and the information that we felt we needed to really inform our decision to make sure we had everything we needed to do the best, in terms of looking at this pipeline and determining whether or not it was in the national interest. So the White House made that decision. We, today, recommended to the President that the permit be denied, and also that he determine that it was not in the national interest. The legislation did not give us enough time to really do responsible evaluation of the factors. And so we continue to believe that this has to be done the right way. So from my perspective, this has not been – this decision is based on what we did November 10th, and then the legislation has not given us adequate time to carry that out.

Question: And my other question was: Do you anticipate another proposal?

Assistant Secretary Jones: Well, the decision that we made – that the President made today accepting our recommendation does – is not made – is based on the fact that we have insufficient information, so it does not preclude other permits of this type that would come in.

Question: All right. Thank you.

Operator: Thank you. The next question we have comes – Neela Banerjee from Los Angeles Times.

Question: Hi. Thank you for holding this call. TransCanada, in its statement today, said that they expect to – that they plan to reapply. And one of the things that they said is that they "expect a new application" – I'm reading from their press statement – "would be processed in an expedited manner to allow for an in-service date of late 2014. TransCanada expects that consideration of a renewed application will make use of the exhaustive record compiled over the past three-plus years."

So I wanted to ask you about that. Do you – I mean, if TransCanada reapplies, as it says it will, what use will the previous assessments – what role will they play in any kind of new EIS? And how realistic is this late 2014 date? Will you have this expedited process as they're – as they say that you will?

Assistant Secretary Jones: Thank you for that question. If the – if TransCanada comes in with a new application, it will trigger a new review process, a completely new review process. We cannot state that anything would be expedited or – at this time. It would just have to go through all of the requirements that are needed for this kind of application review. So I couldn't really speak to when such a review could be finished.

However, I could mention that we do have guidelines that would allow us to use information that's out there. So there is information out there from the process we've been through, but we would also have to look at this as a completely new application, and that's how it would be treated.

Question: Thank you.

Operator: The next question we have comes from Arshad Mohammed from Reuters.

Question: Yeah. It's Arshad Mohammed from Reuters. I just wanted to go back to the last question. If I understood correctly, you said that any application would have to be treated as a new application, so you can't speculate on how long it might take to evaluate. But is it not the case that, both under NEPA and under agency regulations, it is possible for you to take into account the past environmental analyses that have been done? If a new application were to closely track the previous application, then it seems to me that it's only fair to give the company and the broader public some understanding of whether it's going to be soup to nuts, you start everything all over again, or whether it's conceivable that you could save some time if it – if a new proposal didn't differ that much, that you wouldn't have to go over everything that you went over the first time around.

Assistant Secretary Jones: Right. I think – first of all, it's a little bit hard to comment on a permit application that hasn't been submitted yet, and so we haven't seen it. But your point is that we do – and there are regulations under NEPA, and the Department also has internal procedures that outline how to make use of existing NEPA documents, such as the FEIS that we completed in August, when reviewing a new application. So you are correct. The body of information that was out there would inform a new application, but there are certain specific guidelines that have to be used. And beyond that, I can't comment, because it would be a completely new application and a new review process.

Question: So – and when you say it would not be expedited, I guess what you're suggesting is it's not like there would be any fast-track process. But it's – I mean, is it really conceivable that you'd have to go over every single thing anew if it were similar to the existing proposal?

Assistant Secretary Jones: Well, I really – I can't really speculate on that level of detail. I mean, we would treat any future application fairly and in light of all of the information that was brought in with that application, and then we would look at existing information that we have, based on how we're allowed to use that under NEPA procedures and Department guidelines. But that's what we would do. I can't really speak to exactly the steps that you're delineating, because I can't really know exactly what application would be like, and we would have to wait to see what the new application was.

Operator: Thank you. The next question we have comes from Juliet Eilperin, Washington Post.

Question: I may still take – thanks for doing this. I might take one more crack at this to just say that while obviously – could you, first of all, either talk about what you think would have to be updated as part of looking at this application? And again, while you obviously (inaudible) new application, is there anything you can give about how shortened a timeline you could have if you have already factored in the environmental review that obviously has taken several years? I mean, given – if it's a fairly similar route, is there anything you could say about what time process would be involved or, again, things that you could say right now that you feel like, even if they'd been done, the report – the Department would have to look at afresh?

Assistant Secretary Jones: I really can't comment on it, because it is much too speculative. Today, what we're talking about is our recommendations and the decision – the President's decision to accept our recommendation and where we are at this point. I mean, going into the future, I think what I have explained in terms of it being a new process, we would have to follow all of the regulations and guidelines, but we would be able to draw on the information that's out there to the extent it's allowable under the procedures I've already mentioned. But beyond that, I can't really speak to a defined timeline or anything else, because there's no application, and it really is much too speculative.

Question: And then just one quick thing: When – in the statement that the President gave, when he refers to, for example, looking at oil shipments and capacity between Cushing and Gulf Coast refineries, does – would State play any role at all in the decision making on that, or would you be completely removed because it's – again, it's not cross-border and therefore would have actually no intersection with State Department authorities?

Assistant Secretary Jones: Well, an oil pipeline from Cushing, Oklahoma to the Gulf of Mexico would be purely a domestic pipeline, and it wouldn't be crossing any international border, so it wouldn't require a presidential permit, which is the role that the State Department is involved in. So we would not be involved.

Operator: Thank you. The next question we have comes from Mike Holmes, Omaha World Herald.

Question: Dr. Jones, I'm still a little puzzled here. If TransCanada comes back to you with the same route except modifying it to meet its agreement with the state of Nebraska to go around the Ogallala Aquifer, why would you have to review the entire route rather than just the change?

Assistant Secretary Jones: Because it would be – it would be a new application for a permit. And as I've said, we are – what's happening today is the permit that is currently pending is being denied, and so a new permit application would trigger a new process. We would certainly, as I stated, look to the information that's out there to the extent we can, but I – it's a new permit application so the process has to be started over again.

Operator: Thank you. The next question we have comes from Deborah Yedlin, Calgary Herald.

Question: Thanks for taking my call – my question. I'm puzzled as well. I know that the U.S. has made a very big platform of ensuring its energy security. It strikes me that the TransCanada XL pipeline, the Keystone pipeline, would achieve that. So how can you be opining that this is not in the national interest when it goes against this notion of national – of securing your energy future?

Assistant Secretary Jones: The determination that's being made today is based on the fact that we have insufficient information at this point. It goes back to the decision that we made on November 10th when we stated that we needed additional information to really – to really determine if this could be in the national interest. And that is why we cannot at this point go forward, because we don't have that information. It's impossible to get that in the 60-day timeframe that has been spelled out in the act, and so that has led to our actions today.

So the broader issues that you mentioned – certainly energy security, environment, all of those issues – are parts of the national interest, but our decision today is because we have a lack of time to get the information that we think we need.

Question: Thank you.

Operator: The next question we have is Elizabeth Shogren, NPR.

Question: My question follows on that one, which is what is your assessment of the costs that this decision has for energy security?

Assistant Secretary Jones: I really can't speak to that question at this point. I'm sorry, there's some static on the line. I really can't speak to that question at this point. Energy security is one of the considerations in this very complex proposal and in our review process. Today's decision is based on the fact that we do not have adequate information to go forward. And as you know, one of the things required in the act is that there will be a report to Congress, and we will – we think we will need to address some of those issues in that later on today.

Operator: Thank you. The next question we have is from Mike Melia, PBS NewsHour.

Question: Hi there. Thank you for taking the calls. Can you get a little more specific about what were the key points that you felt would not be able to be answered by February 21st? I know you've just said – taking us back a bit to November. But looking at that 60-day timeline, what were the key elements you felt you couldn't do? And we know the President spoke to Prime Minister Harper today. Have you all been in contact with your Canadian counterparts, and what were their reactions?

Assistant Secretary Jones: Okay. In terms of our decision on November 10th, we pointed out that based on the public comments that we had received, we really needed to do in-depth analysis of alternative routes that would avoid the Sandhill area in Nebraska. And there are many dimensions to that analysis and it takes a while to get a lot of that information, and it is something that we just were not able to get that information in terms of how to look at the different perspectives.

For instance, it's my understanding that the applicant has just recently begun to suggest some corridors where alternate routes may be possible, but there has not been much detail work done, which would require a lot of analysis based on the topography and the endangered species and the relative water issues. So much of that information would take quite a while to get. It would take beyond the 60-day timeline.

Question: And a response – have you been in touch with your Canadian counterparts in addition to the President's call?

Assistant Secretary Jones: Today I have not been. I did speak to some – I did reach out to some of my interagency colleagues but not to any of my Canadian colleagues today.

Moderator: All right, Operator, we have time for about two more questions, please.

Operator: Yes, sir. Thank you. The next question is from Andrea Mitchell, NBC News.

Question: Hi, thanks for taking the question. Do you have specific estimates of how much oil you're foregoing? And when you talk about the national interest, what do you say to those who say that energy independence and our reliance on Iran and other unreliable sources in the global market is a much larger threat than this pipeline would be?

Assistant Secretary Jones: Well, I think I would refer you to some of the work that was done in the analysis of the – in the SEIS, where there was analysis in terms of the impact of this pipeline being built or not being built in terms of crude oil coming into the country. And also I think this decision today doesn't make our commitment to energy independence and energy security any less of a priority. It's a major priority for our country. We're making this decision because of the process; we did not have the information we need to make the decision that we thought would be well-informed.

Question: It's still – I guess a lot of us are confused as to how that information could not have been achieved in the weeks that you've had. You were aware of the Nebraska problem for quite some time.

Assistant Secretary Jones: Well, it was on November 10th that we made the decision that we needed the additional information, and since then we have been engaged with discussions with the applicant and with the state of Nebraska. At the end of November, I – or towards the end of November, I believe, the state of Nebraska did pass some legislation regarding rerouting, and we have been working on this. But it is a very – it is a complex issue. Looking for alternative routes and identifying them requires a level of specificity and detail that you have to get into to really be able to analyze them, to compare them. And so it is difficult to get the kind of information that we believe is needed to make a very well-informed decision for the country about this, and we didn't – we knew we could not get that done in the 60-day timeline.

Operator: Thank you. The next question we have is Luiza Savage, MacLeans.

Question: Yes, hello. Thank you for this call. Can you respond to the governor of Nebraska, who had suggested that the State Department issue a conditional permit and allow the state of Nebraska to do its own environmental review of just that alternative route which he said could have been done in the next six to nine months?

Assistant Secretary Jones: Well, it's the responsibility of the State Department to grant this permit which really looks at the crossing of the international boundary, but it looks at the entire pipeline and I think we feel that it would be – it's important for us to look at the whole pipeline and not to really move forward on such a major infrastructure project that will be a part of the country and the landscape for many years in pieces like that. And so I hadn't heard about the governor really proposing this, but we don't really think that's an approach that really deals with the national interest question in an appropriate way.

Question: But this whole Nebraska Sandhills issue, I mean, it was – it's been an issue the whole – for the whole few years of this review, not just since November 10th. It's a little confusing about why it only became an issue for the State Department on November 10th, as you keep pointing out. Why weren't these alternative routes being looked at much earlier in the process so that you would have that information now?

Assistant Secretary Jones: The issue of the pipeline route through the Sandhills and the related aquifers has always been a major concern. It was called out in the SEIS. It led to many discussions. We did look at alternative routes in the SEIS, but they were outside of Nebraska. What we had not done was look at an alternative route that was inside the state of Nebraska.

And in the course of having really extensive public meetings and listening to the citizens of Nebraska and also listening to people across the country, this was a major concern, and we felt that it was important to look at alternative routes within the state of Nebraska that avoided the Sandhills that are really quite a unique terrain. So it's not that this was come to all of a sudden; this has always been something that was of concern. It was addressed in the SEIS. And then during the public meeting session, we received a tremendous amount of comment on this, and it was clear that the importance of the Sandhills was very central to that part of the country.

Moderator: Okay. So I'd like to thank Dr. Jones for taking the time to answer the questions today, and for all of you for participating. Operator, this concludes the call

In: The Keystone XL Pipeline Project
Editors: G. B. Flynn and D. Burnett
ISBN: 978-1- 62081-221-1

Chapter 5

ENERGY ACCESS ACT HEARING—KERRI-ANN JONES TESTIMONY*

Kerri-Ann Jones

Good morning Chairman Whitfield, Ranking Member Rush and other Members of the Subcommittee on Energy and Power; I appreciate the opportunity to appear before you today.

The Keystone XL project is a proposed seven billion dollar, 1,700-mile pipeline that TransCanada Keystone Pipeline LP applied to build and operate. It would transport up to 830,000 barrels per day of crude oil from Alberta, Canada to refineries in the U.S. Gulf Coast Region. Most of the oil would originate from the oil sands of Alberta, although as much as 100,000 barrels per day might have come from the Bakken formation in Montana and North Dakota and up to 150,000 barrels per day could have been transported from Cushing, Oklahoma to the Gulf Coast. The U.S. Department of State received the application for this project in September 2008. We began a thorough, rigorous and transparent process to determine whether issuance of a Presidential Permit for this pipeline was in the national interest. Congress passed the Temporary Payroll Tax Cut Continuation Act of 2011, which required a determination by the President, within 60 days, of whether the

* This is an edited, reformatted and augmented version of the Testimony of Dr. Kerri-Ann Jones, before the Subcommittee on Energy and Power of the Committee on Energy and Commerce United States House of Representatives, dated January 25, 2012.

Keystone XL pipeline project would serve the national interest. On January 18, 2012 the Department of State recommended to the President that the application for a Presidential Permit be denied, due to insufficient time to conduct the necessary analysis, and the President accepted our recommendation and determined that the Keystone XL pipeline project, as presented and analyzed at that time, would not serve the national interest.

I would now like to fill in some further details about our authority to conduct such a process, about the way in which we did so, and about the events that led to our final recommendation. I will also comment briefly on the Administration's view of H.R. 3548, the bill that Congressman Terry and others have introduced seeking to provide an expedited approval of the pipeline by the Federal Energy Regulatory Commission, or FERC.

On April 30, 2004, President George W. Bush issued Executive Order 13337, which amended Executive Order 11423, originally issued in 1968. Executive Order 13337 designated and empowered the Department of State to receive the applications for Presidential Permits for all oil infrastructure projects that cross a U.S. border. The Executive Order indicates that the permit should be granted based on whether it is in the national interest. The Executive Order also states that the Secretary shall consult with eight other agencies: the Departments of Commerce, Defense, Energy, Homeland Security, Interior, Justice, and Transportation, and the Environmental Protection Agency. Executive Order 13337 also envisions broader consultations to include state, tribal and local government officials and foreign governments. The Department's national interest determination factors in a full range of issues, including energy security, foreign policy, economic effects, health, safety, and environmental considerations, including climate change, as well as any other factor the Department believes is relevant to the national interest. To make an informed decision, the Department is directed in the Executive Order to request additional information as needed from the applicant.

In order to analyze the potential environmental impacts of the project, as required by Executive Order 13337, the Department of State determined that it would prepare an Environmental Impact Statement, or EIS, consistent with the National Environmental Policy Act of 1969.

Preparation of an EIS requires significant technical expertise. In a project of this complexity, it requires a great deal of scientific input, including from geologists, hydrologists, ecologists and botanists, to name a few. It requires engineers that understand fluid dynamics and material strength and pipe-welding specifications. It requires statisticians and economists; it requires experts in oil spills and emergency oil spill response. In preparing the EIS for the Keystone pipeline, we consulted extensively with other federal agencies that have substantial direct expertise in these areas.

We followed the provisions of the NEPA regulations, and the example of other federal agencies in engaging a technical contractor to assist in developing the EIS. ENTRIX (now Cardno ENTRIX) was selected. The applicant paid the costs of preparing the EIS. We directed the work of ENTRIX and ensured the quality of all information and analysis in the EIS.

At the same time, we conducted two parallel processes mandated by law. Section 106 of the National Historic Preservation Act of 1966 requires agencies to consider the effects of their actions on properties that may have religious or cultural importance and to engage the Advisory Council on Historic Preservation. In addition, Section 7 of the Endangered Species Act requires Federal agencies to consult with the Fish and Wildlife Service to ensure that the proposed action will not jeopardize the continued existence of any federally listed endangered or threatened species.

Following NEPA requirements, we engaged in a robust public outreach effort. We carried out our consultation on many different levels, including Government-to-Government consultations with over 95 Indian tribes. In 2009, we conducted 20 scoping meetings in communities along the pipeline route. In April of 2010 we issued a Draft EIS and held 21 public comment meetings during a 45-day period including meetings in Kansas, Montana, Nebraska, Oklahoma, South Dakota, Texas, and Washington, DC. As a result of these comments and consultations, we issued a Supplemental Draft EIS in April 2011. In the subsequent 45-day period, we received over 280,000 comments. On August 26, 2011, we issued the Final EIS. Following its issuance we began an interagency review period for the national interest determination required by Executive Order 13337 and conducted an additional public comment period that closed on October 9, 2011. During this period we again held public meetings in Washington, DC and along the pipeline route, including in the Sand Hills of Nebraska.

These meetings were passionate. We heard voices urging diametrically opposite actions and providing solid rationales on both sides. In Nebraska particularly, we again heard concerns, clearly and repeatedly, about the fragile and unique Sand Hills of Nebraska and their importance to the nation and to the people of Nebraska. We listened to these views, many actually supportive of the pipeline, but stressing that the route needed to be moved. Indeed, the people of Nebraska felt so strongly about this issue that their legislators met in special session to draft a law to ensure the Sand Hills would be protected. That is why we paused the process in November 2011, and based on experience with pipelines of similar length we estimated that it would take until early 2013 to complete our assessment.

In December 2011, as we were cooperating with Nebraska's Department of Environmental Quality to develop a process to evaluate any potential new route, the Temporary Payroll Tax Cut Continuation Act (H.R. 3765) was enacted into law imposing a 60-day clock on a decision about the Keystone XL pipeline permit; at that time, the requested permit lacked critical information, including the exact route of the pipeline. We knew that 60 days was not enough time to complete the work and the analysis needed relevant to the national interest determination, especially given that the route had not been finalized. We have been committed to carrying out a thorough, rigorous, inclusive and transparent review of this application and this was not possible within the timeline imposed. We decided – based not on the merits but on the inadequate time period and incomplete review – to recommend that the President deny the permit.

This brings us to H.R. 3548. Last week the President concurred with our recommendation that the Keystone XL project would not serve the national interest at this time. That decision was based on the fact that the exact route of the pipeline has yet to be identified in critical areas. As a result, there are unresolved concerns for a full range of issues, including energy security, foreign policy, economic effects, health, safety, and environmental impacts, among other considerations. This new legislation would not resolve any of these concerns; it just imposes narrow time constraints and creates automatic mandates that prevent an informed decision. The legislation raises serious questions about existing legal authorities, questions the continuing force of much of the federal and all of the state and local environmental and land use management authority over the pipeline, and overrides foreign policy and national security considerations implicated by a cross border permit, which are properly assessed by the State Department.

I know that each and every Member of this Committee is concerned about our nation's energy security, and I can assure you that Secretary Clinton and the Department of State share that concern. Internationally, we remain fully engaged with all our key suppliers, including Canada, as we work on issues of energy diplomacy while at the same time transitioning to cleaner sources of energy. As we do so, the Department of State remains committed to carrying out its responsibilities under Executive Order 13337 with diligence and fairness to the applicants but with ultimate concern for the best interests of the American people.

Thank you again for this opportunity to testify before this Subcommittee and I would be pleased to answer any questions the subcommittee might have.

INDEX

D

E

F

G

H

I

Q

R

S